Multicultural Mathematics Materials

2nd **Edition**

Multicultural
Mathematics
Materials

2nd Edition

Marina C. Krause

California State University, Long Beach

With drawings by the author

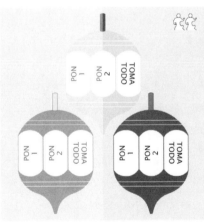

NATIONAL COUNCIL OF TEACHERS OF MATHEMATICS
Reston, Virginia

Copyright © 2000 by
THE NATIONAL COUNCIL OF TEACHERS OF MATHEMATICS, INC.
1906 Association Drive, Reston, Virginia 20191-9988
nctm.org

Library of Congress Cataloging-in-Publication Data

Krause, Marina C.
    Multicultural mathematics materials / Marina C. Krause ; with drawings by the
author.-- 2nd ed.
        p. cm.
    Includes bibliographical references.
    ISBN 0-87353-487-5
    1. Mathematics--Study and teaching--Activity programs. 2. Games in mathematics
education. 3. Multicultural education.

QA16 .K7 2000
372.7--dc21

                                                                                                    00-055447

Printed in the United States of America

Contents

Preface

The activities and games collected here have their roots in different parts of the world. Many have lasted through centuries. Such activities can bring to the present-day mathematics curriculum the vitality of ethnic and cultural diversity. They can further enhance the background of ethnically the "different" child and expose children to the ethnic heritage of others.

The materials are classified by geographic region: Africa, Asia, Oceania, Europe, the Middle East, South America, Middle America, and North America. Certain games, however, are difficult to classify with certainty; evidence of a particular game is sometimes found in more than one part of the world, and time may conceal the actual origin.

The North American region includes a special section devoted to the Hopi Indians of northeastern Arizona. These Hopi materials are the result of a careful search of the literature, firsthand study of artifacts in the Hopi collections of the Southwest Museum in Los Angeles and the Heard Museum in Phoenix, and numerous visits to the Hopi mesas. In my studies I have found that mathematical elements have been woven into the Hopi way of life from its prehistoric beginnings. Those identified here can be incorporated into the curriculum at the elementary school and middle school levels.

The Navajo material in the North American region is expanded. This expansion is the result of numerous visits to the Navajo nation. I found the Navajo Nation Museum and Library in Window Rock, Arizona, a valuable resource in my studies.

Although grade levels are suggested, many of the activities cross grade levels. Depending on the students, some might be used at levels above or below those designated. The game boards and activities are designed for intermediate classroom use. Only simple, readily available materials are required.

I first compiled and developed these materials for multicultural mathematics courses for teachers. Those teachers, in turn, tried the activities in their classrooms. Therefore the activities have withstood the ultimate test: classroom use with children. My hope is that you and your students will find them enjoyable not only from a mathematical but also from a cultural viewpoint.

Egyptian Match

Background

The ancient Egyptian civilization spanned approximately three thousand years. The Pyramids, the oldest buildings in the world today, have already stood nearly five millennia and will be standing millennia from now. The early Egyptian rulers, or pharaohs, were very powerful, the Pyramids were built as tombs to protect their bodies.

Many theories exist about why the pyramid was selected as the shape for these tombs (Weeks 1974). Among them is the practical theory that the pyramid was the easiest way to build a large building. Another theory likened the pyramid to a ramp representing the sun's rays: the dead rulers could climb up this ramp to heaven. These great structures serve as silent testament to the mathematical skills of the Egyptians, who did not have metal measuring tapes; they used measuring cords made of palm or flax fiber.

The Egyptian numeration system was additive—the values of the numerals were added together. They used picture symbols, such as ropes, flowers, and fingers. The symbol arrangement did not matter, and the ropes and flowers could be turned either way. When multiple use of a symbol was involved, the symbols were arranged in sets of three.

Materials

A design or grid that can be partitioned into two matching sides (This design or grid can be drawn on the chalkboard.)

Number of players: 2

Two teams can also be used, thus possibly involving the entire class.

Object

Egyptian match provides practice in both the Egyptian numeration system and bilateral symmetry, or reflection on a line.

How to play

One student or team uses Hindu-Arabic or Roman numerals; the other student or team uses Egyptian numerals. Team 1 selects its side of the design and writes a Hindu-Arabic or Roman numeral in a cell somewhere on the design.

Team 2 must respond with the equivalent Egyptian numeral in the appropriate matching cell on its side of the design. If either condition—the equivalent Egyptian numeral or appropriate place—is not met, Team 1 wins. If Team 2 continues without error until the design is filled, it wins.

After playing Egyptian match

Have students complete the Famous Dates and Pyramid activities.

REFERENCE

Weeks, John. *The Pyramids*. London: Cambridge University Press, 1974.

Solution to Famous Dates

1886
1872
1914
1980
1962
1869
1959

Egyptian Numeration System

Egyptian Numeral	Number Name	Meaning of Picture Symbol
|	1	stroke
∧	10	heel bone
9	100	coiled rope
⚱	1 000	lotus flower
(	10 000	bent finger
🐟	100 000	tadpole
𓀠	1 000 000	astonished person

Famous Dates

Decipher symbols of the ancient Egyptian numeration system to find the famous dates for the historical events given below.

Numeral	Date	Famous Event
𓆼 𓏢𓏢𓏢 𓏢𓏢𓏢 𓏢𓏢 ∧∧∧ ∧∧∧ ∧ ||| |||		Statue of Liberty dedicated on 28 October
𓆼 𓏢𓏢𓏢 𓏢𓏢𓏢 𓏢𓏢 ∧∧∧ ∧∧∧ ∧ ||		First national park—Yellowstone—founded by Congress
𓆼 𓏢𓏢𓏢 𓏢𓏢𓏢 𓏢𓏢𓏢 ∧ ||||		Official opening of the Panama Canal on 15 August
𓆼 𓏢𓏢𓏢 𓏢𓏢𓏢 𓏢𓏢𓏢 ∧∧∧ ∧∧∧ ∧∧		Mt. St. Helens in the state of Washington erupted on 18 May.
𓆼 𓏢𓏢𓏢 𓏢𓏢𓏢 𓏢𓏢𓏢 ∧∧∧ ∧∧∧ ||		Lt. Col. John H. Glenn, Jr., became the first American to orbit the earth on 20 February.
𓆼 𓏢𓏢𓏢 𓏢𓏢𓏢 𓏢𓏢 ∧∧∧ ∧∧∧ ||| ||| |||		Transcontinental railroad completed on 10 May in Promontory, Utah
𓆼 𓏢𓏢𓏢 𓏢𓏢𓏢 𓏢𓏢𓏢 ∧∧∧ ∧∧ ||| ||| |||		Alaska on 3 January became the 49th state, and Hawaii on 21 August became the 50th state.
		The year you were born

Pyramid

If the puzzle instructions are written in Egyptian numerals, then write Hindu-Arabic numerals in the pyramid. If instructions are written in Hindu-Arabic numerals, then write Egyptian numerals in the pyramid.

Across

1. 100
2. 10 011
3. 30
5. ||| / ||| / |||
6. ∩∩∩∩∩ / ∩∩ ||| / ||
7. 1 110 100
9. 120 100
11. 𓏲𓏲𓏲 ∩∩∩ / 𓏲𓏲𓏲 ∩∩ ||| / ||
12. 𓂻𓂻𓂻 / 𓂻𓂻 ∩∩ / ∩∩ |||| / |||
13. 300 000
14. 1 210 020
15. 20 200
16. ∩∩∩ / ∩∩ ||| / ||
17. 11 200
19. 𓂻𓂻𓂻𓂻 𓏲𓏲𓏲 / 𓏲𓏲 ∩ ||
20. 1 011 001
22. 𓏲𓏲𓏲 ∩∩∩ / ∩∩ ||| / ||| / |||
23. 11 101
24. 𓏲𓏲𓏲 ∩∩∩ ||| / 𓏲𓏲𓏲 ∩∩∩ ||| / 𓏲𓏲𓏲 ∩∩∩ |||

Down

1. 121
2. 10 010
4. 1 220 000
5. 𓍢𓍢𓍢 / 𓍢𓍢𓍢 / 𓍢𓍢𓍢 𓂻𓂻𓂻 / 𓂻𓂻 𓏲𓏲 ∩∩ ||| / |||
6. |||| / |||
8. 2 000 000
10. 𓆼𓆼𓆼 / 𓆼𓆼𓆼 / 𓆼𓆼 𓍢𓍢𓍢 𓏲𓏲𓏲 ∩ ||
13. 110 300
15. 10 300
16. 𓏲𓏲𓏲𓏲 ∩∩∩ ||| / 𓏲𓏲𓏲 ∩∩∩ |||
17. 30 000
18. 300
19. 𓏲𓏲𓏲𓏲 ||
20. 1 000 000
21. 10

Pyramid

Pyramid

Senet

Background

Senet, the game of "passing," is an ancient Egyptian game. According to Piccione (1980), senet made its first known appearance in about 2686 B.C. The game was found in the tomb of an Egyptian named Hesy-re, who was an overseer of royal scribes. However, the game may be even older. Game boards of this type dating before this have been found.

From the numerous surviving game sets and the grave sites, Egyptologists know that senet was a popular game with all social classes; kings, nobles, and commoners played senet. Played until the beginning of the Christian era, senet evolved through the centuries from solely a recreational game to a game having both secular and religious forms. The secular form, played by two, continued to be recreational. The other form, played by one, was a ritual game with deeply religious significance.

The funerary senet was placed in tombs for the enjoyment of the dead in the next world. Religious thinkers of the time theorized that perhaps the fate of the dead depended on how well they played senet in the netherworld, or underworld. The dead played against their enemies or perhaps Fate itself to determine whether they would attain immortality. Kendall (1982) maintains that the living most likely played this game because it enabled them to fantasize about the adventures they felt awaited them after death. All this would lead, they hoped, to eternal life in heaven with Ra, the sun god.

The senet board has a path of thirty squares, or "houses." Originally, each player began with seven playing pieces positioned alternately on the first row and squares 11, 12, 13, and 14 of the second row. Square 15 was then the first square open for play.

Changes were made in the game during the Eighteenth Dynasty, 1570-1293 B.C. The number of playing pieces decreased to five. Gaming boxes were made, with senet on one side and a game called twenty squares on the reverse. A drawer in the middle of the box held playing pieces and casting sticks. An exquisite ebony, ivory, and gold game board of this type was found in King Tutankhamen's tomb.

The boards were decorated in a variety of ways. Early recreational boards were marked simply; later ritual boards pictured gods. Square 15 and the last five squares in row 3 seem to have had significance on all boards.

Although detailed rules of play are unknown (Grunfeld 1975), Egyptologists have provided information that gives a better understanding of the game. Consequently, senet can be played today.

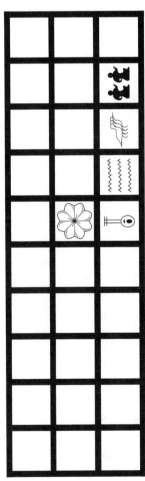

7

Materials

- 3 × 10 senet board
- 5 gold playing pieces
- 5 silver playing pieces
- 4 casting sticks—black on one side and white on the other. Half-round dowels or Popsicle sticks can be used.

Number of players: 2

Object

Senet is a game of logic involving counting. Players attempt to pass the opponent and remove all their own playing pieces safely from the board. It is also a game of strategy. Players can develop strategies to block opponents and send them backward on the board. Occupying "good" squares and using blocking maneuvers can force an opponent to undesirable locations. Since moves are based on the toss of casting sticks, chance is also an element. Older students can determine why particular tosses are easy or difficult to obtain.

How to play

Prepare the board by arranging the ten playing pieces on the top row, left to right, in the following order: silver, gold, silver, gold, and so forth. Playing pieces are moved from left to right along the top row, right to left along the middle row, and left to right along the bottom row:

1	2	3	4	5	6	7	8	9	10
20	19	18	17	16	15	14	13	12	11
21	22	23	24	25	26	27	28	29	30

Interpret the tosses of the casting sticks as follows:

> One white side up equals 1
> Two white sides equals 2
> Three white sides equals 3
> Four white sides equals 4
> Four black sides equals 6

A toss of 1, 4, or 6 entitles the player to keep playing. After a toss of 2 or 3 and its move, play goes to the other player.

Interpret square 15 and the last five squares as follows:

- **Square 15:** "House of Repeating Life," a good square. If a playing piece lands here by exact count, a player is entitled to a free turn.
- **Square 26:** "House of Rejuvenation," a good square. If a playing piece lands here by exact count, the player is entitled to a free turn. From square 26, players try to cross over square 27, a definite pitfall.
- **Square 27:** "House of Humiliation," "House of Misfortune," or "Waters of Chaos." As the names imply, square 27 is a bad square. If a playing piece lands here by exact count, the player must make a choice: The playing piece must either be returned to square 15 or remain on square 27 with no movement of any of the player's pieces until a 4 is tossed. A 4 removes the piece from the board. If square 15 is occupied, the piece is placed on the empty square closest to square 15.
- **Square 28:** A good square, since the playing piece has successfully crossed over the "Waters of Chaos." A 3 must be tossed to remove the piece from the board.
- **Square 29:** A good square. A 2 must be tossed to remove the piece from the board.
- **Square 30:** A good square, since it is the square of Re-Horakhty (the sun god's name as he rises into the dawn). In the ritual game, removing the pieces from the board signifies the deceased's journey out of the netherworld and union with the sun god. A 1 must be tossed to remove the piece from the board.

Squares 15, 26, 28, 29, and 30 are "safe havens." Playing pieces on these squares are free from attack.

Gold plays first. Casting sticks are thrown to determine who plays gold, and the highest toss wins. Players take turns. A toss of the casting sticks can be used to move any *one* of a player's pieces. Note that if a 4 is tossed, the player cannot decide to move one piece one square and then a second piece three squares.

Two playing pieces of the same color cannot occupy one square. An opponent's playing piece is attacked by landing on the same square during a forward move. The opponent's playing piece is then moved back to the square left by the attacking piece.

When a forward move cannot be made, the turn must be used to move backward. Landing in reverse on a square occupied by an opponent moves the opponent to the forward square just vacated. If a move forward or backward is not possible, the turn is forfeited.

Playing pieces of the same color in two or three consecutive squares are safe from attack. Three playing pieces of the same color on three consecutive squares constitute a blockade. No other playing pieces can pass, but they can move backward if movement in that direction is possible.

To remove playing pieces from the board, players must make an *exact* toss of 4, 3, 2, or 1 when the pieces are in squares 27, 28, 29, or 30. If the number tossed is greater than the exact number needed to move a playing piece off the board, the player can either move another piece or move backward. If no forward or backward moves are possible, the player forfeits the turn. The player who gets all his or her playing pieces off the board first wins.

REFERENCES

Grunfeld, Frederic V., ed. *Games of the World.* New York: Holt, Rinehart & Winston, 1975.

Kendall, Timothy. "Games." In *Egypt's Golden Age: The Art of Living in the New Kingdom, 1558–1085 B.C.,* edited by Edward Brovarski, Susan Doll, and Rita Freed, pp. 263–70. Boston: Museum of Fine Arts, 1982.

Piccione, Peter A. "In Search of the Meaning of Senet." *Archaeology* 33 (July/August 1980): 55–58.

Twenty Squares

Background

Twenty squares, found on the reverse of gaming boxes with senet, remains a mystery. Little evidence is available concerning how the Egyptians played the game. The arrangement of the squares is similar to that of senet because three rows or ranks of squares were used; however, the middle row consisted of 12 squares with 4 squares at each end in the rows above and below (see fig. 1). The game was played by two people, probably with one astragal (anklebones) as a die or with a teetotum, a four-sided top with sides labeled 1, 2, 3, and 4. According to Kendall (1982), some existing boards appear unmarked, but most have every fourth square marked with a rosette or good-luck symbol. In some cases, words or phrases have been inscribed on certain board squares to indicate an advantage or disadvantage for landing there. It is not known how the playing pieces were placed or brought on the board to begin the game. One may speculate that players started at opposite ends of the board and, somehow moving in opposing directions, made their way through the middle row and off the board.

Materials

twenty squares board
5 gold playing pieces
5 silver playing pieces
spinner with 4 divisions

Number of Players: 2

Object of the Game

It seems likely from the observable evidence—existing playing boards—that twenty squares, like senet, is a game of logic involving counting. To play twenty squares, students can suggest and create a set of instructions for playing.

Activity

1. Distribute a twenty squares board, playing pieces, and a spinner.
2. Ask students to study the arrangement of the squares and the markings on the board.
3. Have them consider if these markings might have been "safe havens," what advantage or disadvantage might be involved if a player lands on a certain square, how a player might blockade an opponent, and how a player might leave the board and win.
4. Play twenty squares.
5. Ask if the instructions for playing work. If not, how could they be changed?

Twenty Squares

Fig. 1.

Wari

Background

The Arabic word *mancala,* meaning "transferring," is the generic name for a set of similar board games played in a number of different regions of the world. Although the names, rules, and boards vary, the principles remain basically the same. Some games are played on boards with two rows of six holes each; other, more complex games are played on boards with four rows of eight holes each.

Mancala games have been known and played for thousands of years. In Egypt the boards have been discovered carved into the stone of the great pyramid of Cheops and the temples at Luxor and Karnak. The game spread to the rest of Africa and to Asia, the Philippines, the West Indies, and Surinam in South America as well.

Wari, also known as oware, is a two-row version of the game played by the Ashanti people of Ghana in West Africa. Located on the Gulf of Guinea, Ghana is predominantly agricultural; the biggest crop and major export is cacao (chocolate).

Materials

1 egg carton that will hold 12 eggs

48 beans or pebbles

Remove the lid of the egg carton and cut it in half. Attach one half of the lid to each end of the lower portion of the carton.

Number of Players: 2

Object of the Game

Wari is a game of logic—each player must consider the advantages and disadvantages of possible moves. For young students wari provides practice in counting. The object is to capture 25 beans, more than half the total. (If desired, the player having the most beans at the end of the game can be considered the winner.) Players must pay attention not only to their own side of the board but also to their opponent's side. Cups having one or two beans are vulnerable to capture. Players can defend their cups by moving their own beans so that one or two do not remain in a single cup. Or they can move their beans around the board to load the opponent's cups in such a way that the opponent's last bean will not land in the other player's vulnerable cups.

How to play

To begin, four beans are placed in each egg cup. Players face each other with the board between them. A player's side of the board includes the six cups on his or her side and the lid, or pot, to the right. Movement is to the player's right—counterclockwise.

The **first player** removes four beans from any cup on his or her side and sows them one by one in the next four cups. On the first move it is possible, depending on which cup is emptied, to place beans in the opponent's cups. (Players do not sow beans in either of the pots; these are for captured beans only.) The **second player** removes all the beans from any cup on his or her side and distributes them similarly around the board.

Since cups can hold many beans, it is possible to go around the board more than once during a turn. As players go around the board, they skip the empty cup from which they last removed their beans and leave it empty for that turn.

12

Captures are made as follows:

1. If the last bean a player sows lands in a cup on the opponent's side that contains only one or two beans, that cup is then captured. This last bean plus the opponent's bean(s) must total no more than three.

2. When this condition is met, the player is also permitted to take beans from all cups preceding it if (*a*) each cup contains two or three beans, (*b*) the cups are on the opponent's side of the board, and (*c*) the cups are consecutive.

All captured beans as well as the last bean are placed in the player's pot.

No move is permitted that would enable the player to capture all the opponent's beans. Such a move would make it impossible for the oppo-nent to play. When one side of the board is empty and it is the opposing player's turn, that player must move beans to the other side if possible so the game can continue. If, however, it is the player's side that is empty, then the game is over. The player who still has beans adds them to his or her own pot.

If a game continues on and on with only a few beans on the board, players may agree to stop. They then take the beans on their side of the board for their own pot, and the player with the most beans wins.

ADDITIONAL READING

Grunfeld, Frederic V., ed. *Games of the World*. New York: Holt, Rine-hart & Winston, 1975.

Zaslavsky, Claudia. *Africa Counts*. Boston: Prindle, Weber & Schmidt, 1973.

Ko-no

Background

Ko-no comes from China and Korea. *Ko-no* is the Korean name for games played on a diagram, and *k'i* is the Chinese name. The playing pieces are black and white stones.

Materials

game board
2 black playing pieces
2 white playing pieces

Number of players: 2

Object

Ko-no is a game of logic. The object is to block the opponent so that he or she cannot move.

How to play

One player uses the black pieces and the other uses the white ones. The black pieces, placed at *A* and *B*, are played first. The white pieces are placed at *C* and *D*. Players take turns by moving one playing piece along any segment to an unoccupied position. No jumps are allowed. The game is over when a player is blocked and cannot move either playing piece.

Discuss the winning and losing plays for ko-no. If both players play logically, what happens?

ADDITIONAL READING

Culin, Stewart, *Games of the Orient: Korea, China, Japan.* 1895. Reprint. Rutland, Vt.: Charles E. Tuttle Co., 1958.

Ko-no

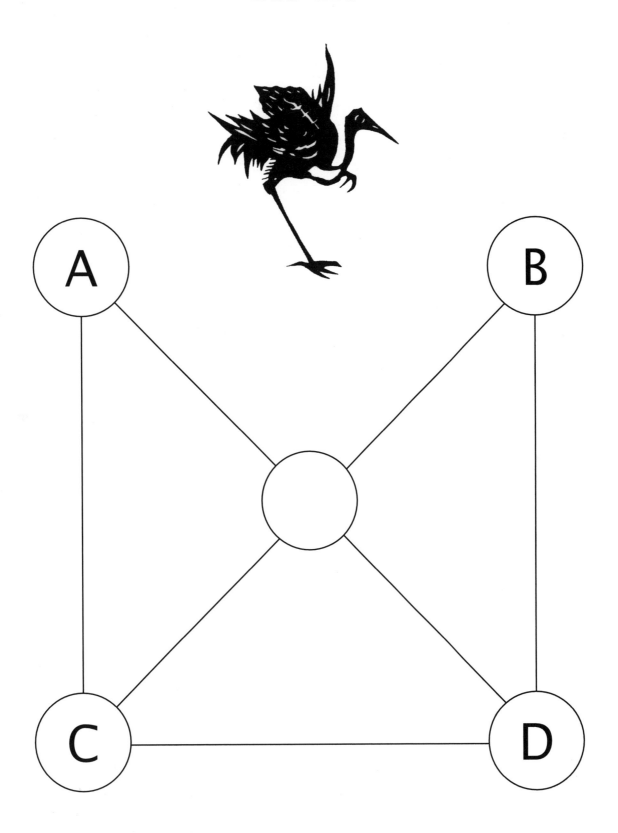

Five-Field Ko-no

Background

In Korea, *ko-no* is the name for games played on diagrams. In addition to the ko-no game presented earlier, Koreans also play four-, five-, and six-field ko-no. Interestingly, five-field ko-no, called *o-pat-ko-no* in Korean, is played on a 4 × 4 diagram. Notice that a 4 × 4 diagram has *five* line segment intersections **across** (five rows) and five **down** (five columns). Playing pieces are moved along these lines.

Materials

game board
7 small black stones or playing pieces
7 small white stones or playing pieces

Number of players: 2

Object

To occupy the opponent's side of the board in the starting arrangement before the opponent occupies your side.

How to play

Five-field ko-no is a strategy game. Arrange the playing pieces on the game board as shown. Black playing pieces move first. Moves are made along the line segments. A turn consists of moving one square forward, backward, or diagonally across a square, even though diagonals are not drawn on the board. Players take turns. Jumping is not allowed. The winner is the first player to get his or her playing pieces to the opposite side of the board in place of those of the opponent.

ADDITIONAL READING

Culin, Stewart. *Games of the Orient: Korea, China, Japan.* 1895. Reprint. Rutland, Vt.: Charles E. Tuttle Co., 1958.

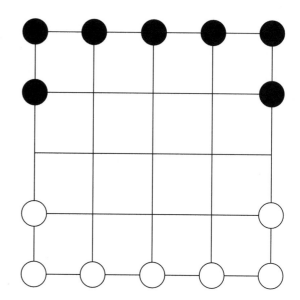

Five-Field Ko-no

Nyout

Background

Nyout is a popular race game from Korea. The gameboard is a combination of a circle and a cross. Centuries old, nyout comes to the present unchanged. Young and old enjoy playing this game.

Materials

game board

4 casting sticks, white on the flat side and black on the convex side (Painted Popsicle sticks or half-round, painted dowels could be used. Some traditional games are played with casting sticks similar to blocks 2.5 cm long; other games use sticks about 20 cm long.)

4 playing pieces, known as *mal* or "horses," per player

Number of players: 2–4

If two play, then each tries to get one or four horses around and off the board. Before play begins, players must agree to the number of horses. If three play, each tries to get three horses around and off the board. If four play, each tries to get four horses around and off the board. Also when four play, those sitting across from each other are considered partners (Culin [1895] 1991).

Object

Nyout is a race game of chance involving casting sticks used as dice. Players attempt to move one to four horses around and off the board before others do. Players can develop strategies to capture their opponents' horses and send the captured horses back to begin again.

How to play

The order of play is determined by each player tossing the casting sticks. Highest plays first and descending order is used. Tosses of the casting sticks are interpreted as follows:

> One *(to)* white side up equals 1
> Two *(kăi)* white sides up equals 2
> Three *(kel)* white sides up equals 3
> Four *(nyout)* white sides up equals 4
> Four *(mo)* black sides up equals 5

A *nyout* (toss of 4 white or flat sides, which equals 4) or a *mo* (4 black or round sides, or 5) gives the player another toss. The tosses may be divided between two horses belonging to the player. When four players play, a player may move his partner's horse.

To begin the game, the first player tosses the casting sticks and enters the board on the small circle to the left of the exit. Movement is counterclockwise to the exit. If a horse lands on one of the four large outer circles (see A–D) by exact count, a shortcut may be taken along the inner cross of the board to the exit. For example, if a player tosses a *mo*, or 5, in the beginning, the horse enters the board and lands on B. Next movement of the horse may be from B to E.

When a player's horse lands on another of his own, the horses may be moved together as one on the board. When they exit the board, they are counted as two.

When a player's horse lands on an opponent's horse, the opponent's horse is captured and must be taken off the board to begin again. After a capture is made the player making the capture gets another toss.

To remove a horse from the board, a toss of one more than the number of moves to the exit circle is required. For example, if a player is 2 from the exit (A), then a *kel*, or 3, must be tossed.

REFERENCE

Culin, Stewart. *Korean Games with Notes on the Corresponding Games of China and Japan.* 1895. Reprint. New York: Dover Publications, 1991.

Nyout

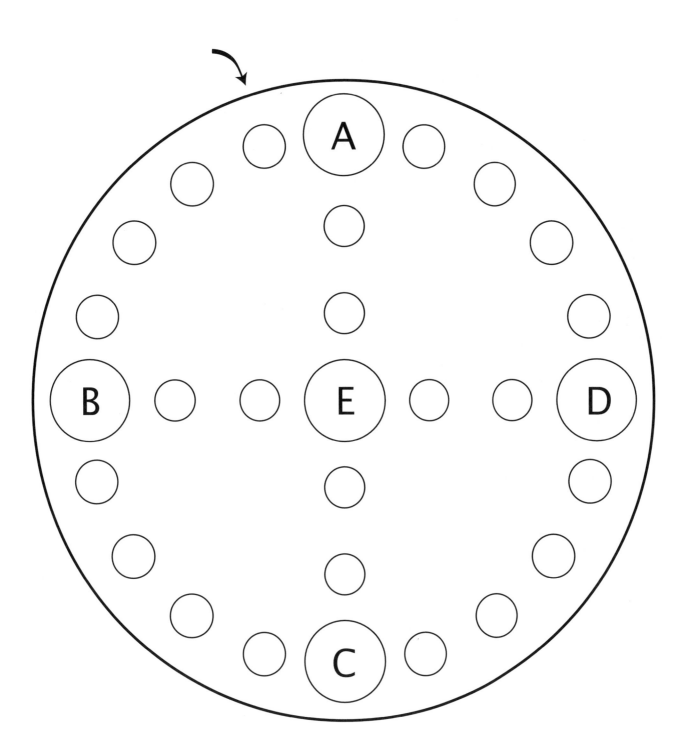

Tangram

Background

The tangram comes from China. Its origin is not known, and this adds an air of mystery to the puzzle. Many believe that the tangram must be, like other things from China, very old. Tangrams became known to the Western world during the nineteenth century (Read 1965). Sometimes people can be heard calling other geometric puzzles "tangrams," but the true Chinese tangram consists of seven special shapes: two large congruent triangles, one medium-sized triangle, two small congruent triangles, one square, and one parallelogram. These shapes will form—in addition to a square—a rectangle, a parallelogram, a trapezoid, and a triangle. Besides these geometric shapes, an infinite number of designs can also be formed.

Materials

Scissors for each student 1 square of construction paper for each student, at least 15 cm × 15 cm

Object

Constructing and manipulating the pieces, or "tans," of the tangram offer many opportunities for students to make geometrical discoveries concerning shapes, congruent triangles, and area. Students may create their own designs as well.

Activity

Constructing the tangram pieces

1. Fold the square in figure 1 in half diagonally and cut on the fold. Two large congruent triangles result.
2. Fold one of the triangles in half and cut on the fold (fig. 1). Two triangles are formed.

Save these two pieces. Each of these two pieces is 1/4 of the original square.

3. Take the other large triangle (△ABC) and fold B to the midpoint of $\overline{AC}$ and cut off the triangle formed (fig. 2). Save this triangular piece. This triangular piece is 1/8 of the original square.

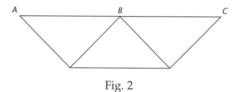

Fig. 2

4. The shape that remains is shown in figure 3. Fold the shape in half. C and Y are now on top of A and X (fig. 4).

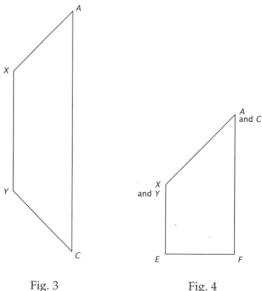

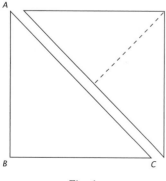

Fig. 1

Fig. 3

Fig. 4

5. Unfold, and fold *C* to *F* (fig. 5).

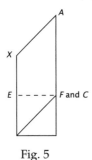

Fig. 5

6. Unfold and fold *X* down to *F* (fig. 6).

7. The shape now has these three folds in it (fig. 7). Cut on the folds and save the four shapes. The puzzle pieces thus formed are two triangles, one square, and one parallelogram. Each small triangle is 1/16 of the original square. The square and parallelogram are each 1/8 of the original square.

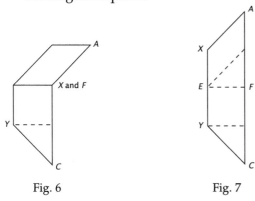

Fig. 6   Fig. 7

Have students consider the seven shapes found: five triangles, one square, and one parallelogram. They will find that the two large triangles and the two small triangles are congruent. Students may make other discoveries. The two small triangles will fit on top of the medium-sized triangle, the square, or the parallelogram.

Activities

1. Have students put the square together again (fig. 8).

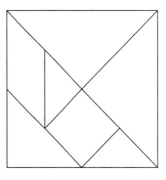

Fig. 8

2. Ask students to remove the two largest triangles from the solution. By positioning the two largest triangles around the remaining half of the solution, a rectangle, a parallelogram, a trapezoid, and a triangle can be formed.

3. Duplicate and distribute the cat puzzles that appear on the following pages. Younger students may match shapes, or tans, to each puzzle picture.

4. Trace the outlines of the cats. Older students may position the tans so that they fit within the outline.

REFERENCE

Read, Ronald C. *Tangrams—330 Puzzles*. New York: Dover Publications, 1965.

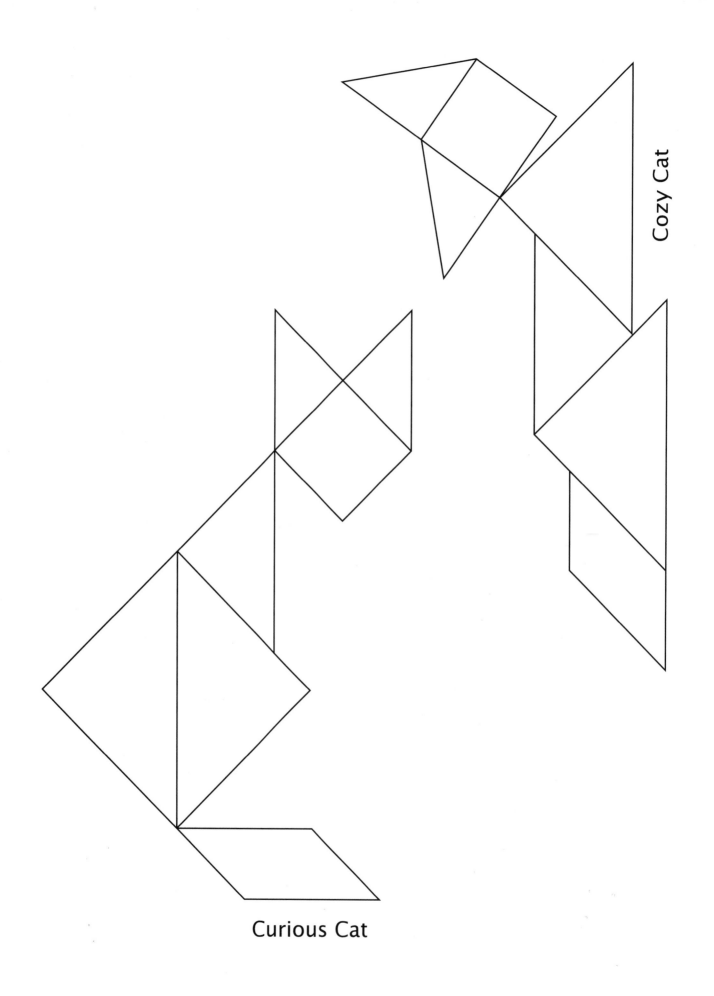

Cozy Cat

Curious Cat

Suspicious Cat

Dancing Cat

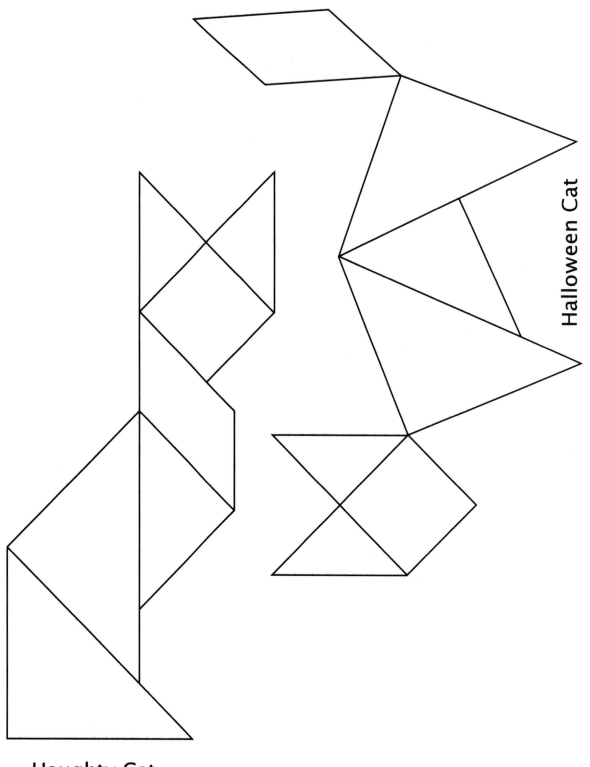

Halloween Cat

Haughty Cat

Galosh

High Heel

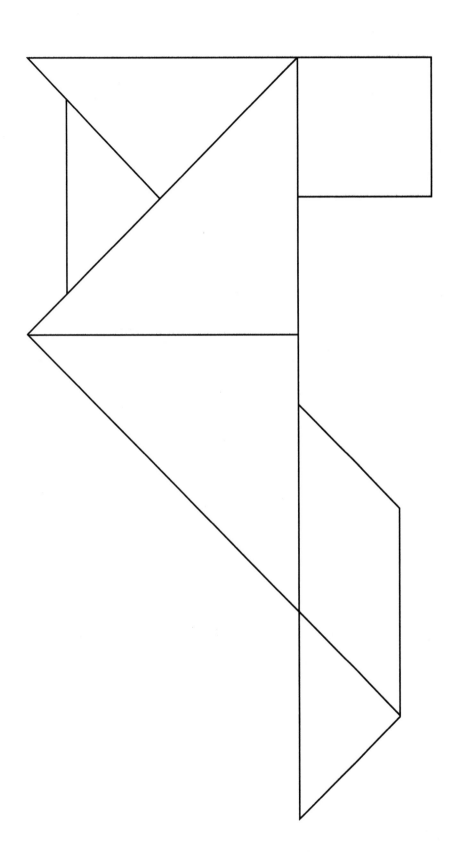

Pump

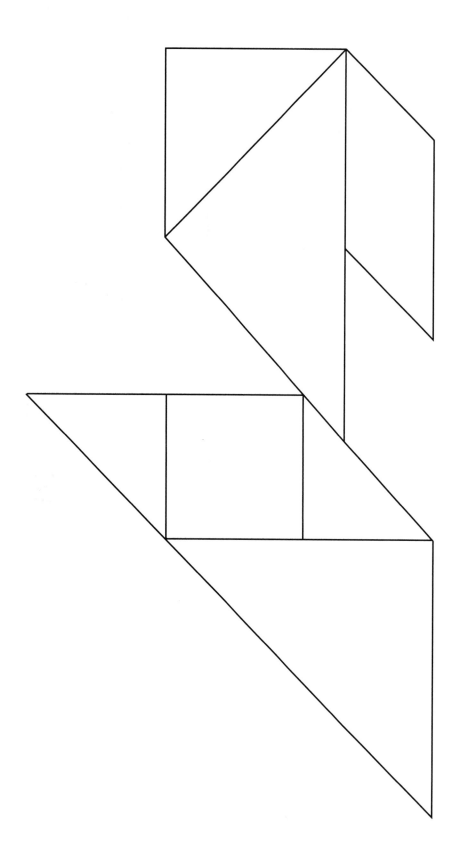

Slipper

Tap Shoe

Clog

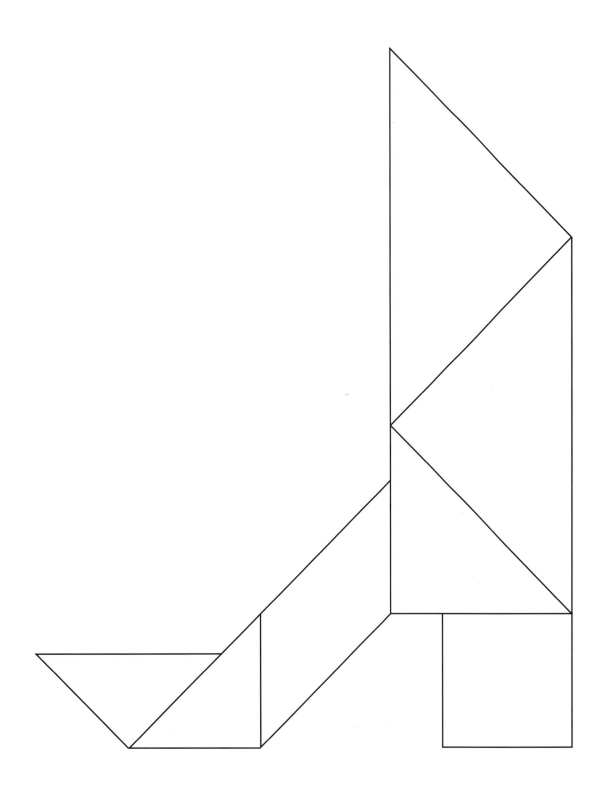

Boot

Origami Cup

Background

Origami, a Japanese word, means "the folding of paper." The origins of origami are unknown. In A.D. 538, a Buddhist priest brought methods of paper making to Japan from China and Korea. Logic dictates that methods of paper folding came in at the same time, although historians are unable to document this.

Since paper was expensive, ancient origami was used only for ceremonial occasions. The ancient ceremonial styles of origami were difficult, and to learn them people had to be taught by a specialist. Consequently, the ceremonial styles were not generally known. In the nineteenth century, after contacts with the West, the Japanese dropped the ancient origami styles in favor of modern works, those that are known today.

Origami is a handicraft enjoyed by both adults and children. Japanese mothers pass on paper-folding methods to their children.

Materials

Paper squares 18 cm × 18 cm. Size of square and type of paper can vary. Since a drinking cup is being made, waxed paper can be used.

Object

Students will construct a paper cup that can be used. The folding activity is geometrical. Geo-metric graphic aids and symbols are given in the instructions. This activity also provides practice in following instructions, which can be given orally, with a few illustrations on the chalkboard, or in written form.

1. Fold a square in half diagonally from corner to corner (fig. 1).
2. Fold forward on $\overline{FG}$ (fig. 2), with C going to D (fig. 3).

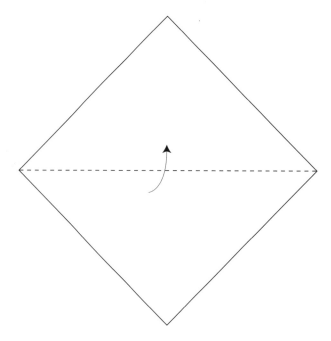

Fig. 1

32

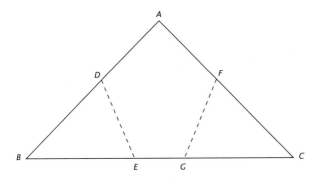

Fig. 2

3. Fold backward—away from your body—on $\overline{DE}$; B goes to the back of F (fig. 4).

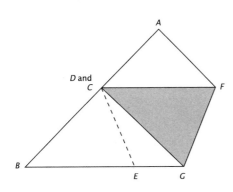

Fig. 3

4. Fold $\triangle ADF$ toward you and down into the paper slot (fig. 5).

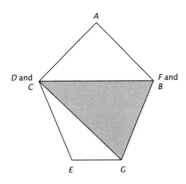

Fig. 4

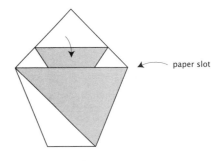

paper slot

Fig. 5

5. Fold the remaining top triangle down into the slot on the other side (fig. 6). The cup is now ready to use.

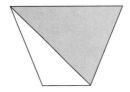

Fig. 6

ADDITIONAL READING

Honda, Isao. *The World of Oragami.* 1965. Reprint. Tokyo: Japan Publications Trading Co., 1976.

Magic Squares

Background

A magic square is an arrangement of numbers in the form of a square so that when the numbers are added vertically, horizontally, or diagonally, the sum will be the same. Known from very ancient times, magic squares have been discovered in such widely divergent cultures as China, Egypt, India, and western Europe. It was believed that they had magic properties and perhaps were even connected with the stars. According to a Chinese myth, the magic square, called *lo-shu*, appeared on the back of a divine tortoise in the Yellow River about 2200 B.C.

Materials

magic square grids, one or more for each student

Object

Magic squares provides practice in addition and subtraction. By following three rules, students can apply the pattern of movement to construct additional magic squares.

Activity

To construct magic squares for odd numbers squared ($3^2, 5^2, 7^2, \dots$), follow these three rules:

1. Position the numerals in consecutive order, beginning with 1. Place the numeral 1 in the top center cell.
2. Proceed diagonally upward and to the right from each small square. If you leave the large square at the top, drop to the bottom of the column. If you leave the large square at the side, go to the other end of the row.
3. If a number is a multiple of the number that is squared to get the total number of cells in the magic square, the next numeral is placed directly below. Figure 1 shows the pattern of movement for the construction of a (3^2) magic square.

Have students experiment with making magic squares using the set of natural numbers—the set of numbers beginning with 1 and having no

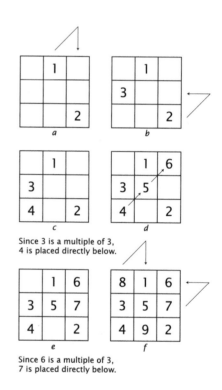

Since 3 is a multiple of 3, 4 is placed directly below.

Since 6 is a multiple of 3, 7 is placed directly below.

Fig. 1

largest member. This pattern of movement and square form works not only for consecutive numbers but for consecutive odd or even numbers as well. See figures 2 and 3.

19	5	15
9	13	17
11	21	7

Fig. 2

20	6	16
10	14	18
12	22	8

Fig. 3

Magic Squares

17		1	8	15
	5	7	14	
4	6	13		22
10	12	19		3
11	18		2	

Sum: _____

Sum: _____

34		2	16	30
46		14	28	
8	12		40	44
20	24			6
22		50	4	18

Magic Squares

30	39		1	10	19	
		7	9		27	29
46	6	8		26		37
5	14	16	25	34	36	
13	15	24	33	42	44	4
21	23			43	3	12
	31	40	49	2	11	20

Sum: _____

Make your own

Sum: _____

Sum: _____

Lu-lu

Background

Polynesian voyagers were the earliest known settlers of the Hawaiian Islands. The English explorer and navigator Captain James Cook discovered the islands and landed at Waimea, Kauai, on 20 January 1778. He named them the Sandwich Islands after John Montagu, the earl of Sandwich.

The islands are of volcanic origin. The Hawaiians played lu-lu with disks of volcanic stone about 2.5 cm in diameter. These disks served as stone dice called *u-lu*. The markings on the stone dice were painted red. The word *lu-lu* means "to shake."

Materials

Four playing pieces marked on one side only as follows:

Playing pieces can be constructed from such materials as clay, shell, or wood. Buttons can also be used—masking tape can be affixed to them and marked with a felt-tip pen.

Number of players: 2 or more

Object

Lu-lu is an uncomplicated counting game.

How to play

A player shakes the playing pieces in both hands and tosses them. A turn consists of two tosses:

- If all four playing pieces fall face up, the player scores ten and then tosses all pieces for a second time.
- If all four playing pieces do not fall face up, the player counts and scores the dots of those face up and then tosses only the face-down pieces a second time. The dots that show on the second toss are counted with those of the first toss.

The winner can be the player with the highest score in a single turn, or play can continue to an agreed-on score, such as 100.

ADDITIONAL READING

Culin, Stewart. "Hawaiian Games." *American Anthropologist* 1 (April 1899): 201–47.

Konane

Background

Konane (pronounced ko-nah-nay), the Hawaiian checkers game, was played by the Hawaiians before the first western explorers came to the islands. It is said that Kamehameha the Great, king of the island of Hawaii after 1790 and ruler of all the Hawaiian Islands from 1810 until his death, was a master of the game. Playing boards made from wood as well as stone have been recovered from ancient village sites.

Materials

A game board, called a *papamu*. On old game boards, the number of positions varied from 64 to 260. For the classroom, mark 64 positions—8 rows of 8 dots—on a 30 cm × 30 cm piece of posterboard or cardboard.

The game pebbles, or *ʻiliʻili*, are as follows:
32 white pebbles, or *ʻiliʻili keʻokeʻo*
32 black pebbles, or *ʻiliʻili ʻeleʻele*

Pieces of water-worn coral were used for white, and black volcanic stones were used for black.

Number of players: 2

Object

The object of konane is to maneuver opponents to a position from which they can no longer move any of their playing pieces.

How to play

Players cover the entire game board alternately with black and white playing pieces (fig. 1). To determine who begins, one player removes two adjacent playing pieces, one white and one black, from near the center of the board and conceals one in each hand. The other player chooses a hand, and the color of that playing piece is the choosing player's color. Black plays first. The two pieces are not returned to the papamu, or game board.

All moves are made by jumping. A jump consists of going over an opponent's playing piece into a vacant place and removing the opponent's piece from the papamu. Multiple jumps on one turn are permitted.

Moves may be in a horizontal or vertical direction, never diagonally. Movement in more than one direction in a single turn is not permitted. It is permissible, however, for a player to decide not to take all the available jumps in a single turn.

With fewer and fewer playing pieces on the game board, fewer and fewer opportunities exist for jumping. The first player who is unable to jump over an opponent's playing piece loses the game.

ADDITIONAL READING

Buck, Peter H. (Te Rangi Hiroa). *Arts and Crafts of Hawaii*. Section VIII, "Games and Recreation." Bernice P. Bishop Museum Special Publication no. 45, 1957. Reprinted in Separate Sections. Honolulu: Bishop Museum Press, 1964.

Mitchell, Donald D. *Hawaiian Games for Today*. Honolulu: Kamehameha Schools Press, 1975.

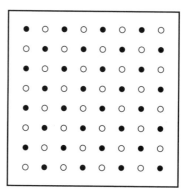

Fig. 1. Konane board layout

Petroglyph

Background

A petroglyph is a picture carved on a rock. Petroglyphs have been found in various parts of the world where early peoples lived. The petroglyph for this activity comes from Hawaii.

Little is known about the origin of the petroglyphs found in Hawaii. This is not surprising; historical evidence is usually not available for the origins of petroglyphs found in other parts of the world either. It is not unreasonable to assume that the Hawaiian petroglyphs were carved by the ancestors of the Hawaiians living today. The Hawaiians themselves, however, do not know who carved them or why.

The petroglyph for this activity represents a muscled figure. According to Cox and Stasack (1980), the portrayal of the human image as a triangular, muscled figure distinguished Hawaiian petroglyphs from petroglyphs found in other places. In addition to triangles, Hawaiian petroglyphs have circles, concentric circles, dots connected to curving lines, and U shapes. Interestingly, rectilinear forms were rarely used by the makers of the petroglyphs.

Materials

petroglyph drawing
paper
pencil or marker

Object of the Activity

The students draw a Hawaiian petroglyph of a muscled figure. In doing so, students are drawing pictures of triangles, simple closed figures having three sides. Older students can classify the triangles by their sides or angles.

Activity

Ask students if they know what a petroglyph is. If no one knows, discuss how they can find out. After a definition has been obtained, introduce Hawaiian petroglyphs using the background information and petroglyph drawing provided.

Pose the following questions:

1. What shapes do you see in this petroglyph?

2. Do you think this petroglyph would—or would not—be easy to draw? Why?

Try it and see.

After drawing the petroglyph, older students can classify by sides the triangles they have drawn. That is, is a triangle isosceles, equilateral, or scalene? An isosceles triangle has two sides equal. An equilateral triangle has all three sides equal, whereas all three sides of a scalene triangle are unequal.

Older students can also classify the triangles they have drawn by angles. That is, is a triangle acute, right, or obtuse? An acute triangle has every angle measuring less than a right angle. A right triangle has one right angle. An obtuse triangle has one angle that is greater than a right angle.

REFERENCE

Cox, J. Halley, and Edward Stasack. *Hawaiian Petroglyphs.* Bernice P. Bishop Museum Special Publication no. 60. Honolulu: Bishop Museum Press, 1970.

Petroglyph

Tapatan

Background

Tapatan is a Philippine game that resembles games from other cultures, some of them ancient. A game board similar to that of tapatan is one of seven cut into roofing slabs at the temple of Karnak, Egypt; it thus dates from 1400 to 1333 B.C., when the temple was constructed. The Chinese played the same game, which they called *luk tsut k'i,* "six-man chess," around 500 B.C., during the time of Confucius.

According to Bell (1979), the Roman poet Ovid (43 B.C.–A.D. 18) refers to the game in his *Ars amatoria (The Art of Love).* Another version, three men's morris, was popular around 1300 in England, where boards have been found cut into cloister seats in great cathedrals such as Canterbury and Westminster Abbey.

A hardwood tapatan board, acquired in 1892, was part of a collection made by Alexander R. Webb, then U.S. consul at Manila, Philippines, for the United States National Museum, now part of the Smithsonian Institution in Washington, D.C. (Culin 1900). The board measured 10.5 inches, or about 26 centimeters, square. The playing pieces were six round pieces of wood, three white and three dark. Webb reported that many Philippine families liked and played tapatan; diagrams of the game's board were often drawn on the doorsteps or floors of native houses.

Materials

tapatan board
3 white playing pieces
3 dark playing pieces

Number of players: 2

Object

A player's objective is to get three playing pieces of the same color in a line. Tapatan is played on a geometric field. Playing pieces can be placed or moved on nine points. The first player has three choices: the center, the middle of a side, or a corner. Note that the first player has the advantage and can force a win. After students discover which player can always win if no mistake is made, a spinner or die can be used to determine who plays first

How to Play

Dark plays first. The first player begins by placing one playing piece either in the center of the board where the line segments intersect or where any one of the line segments touches a side or corner of the square. The second player then places a playing piece on the board. Players take turns. After all six playing pieces are on the board, play continues with each player sliding a playing piece along a line to another position. Jumps are not allowed. The winner is the first player to get three playing pieces of the same color in a row, a column, or a diagonal.

REFERENCES

Bell, Robert C. *Board and Table Games from Many Civilizations.* 1960. Reprint. New York: Dover Publications, 1979.

Culin, Stewart. "Philippine Games." *American Anthropologist* 2 (October–December, 1900), 643–56.

McConville, Robert. *The History of Board Games.* Palo Alto, Calif.: Creative Publications, 1974.

Tapatan

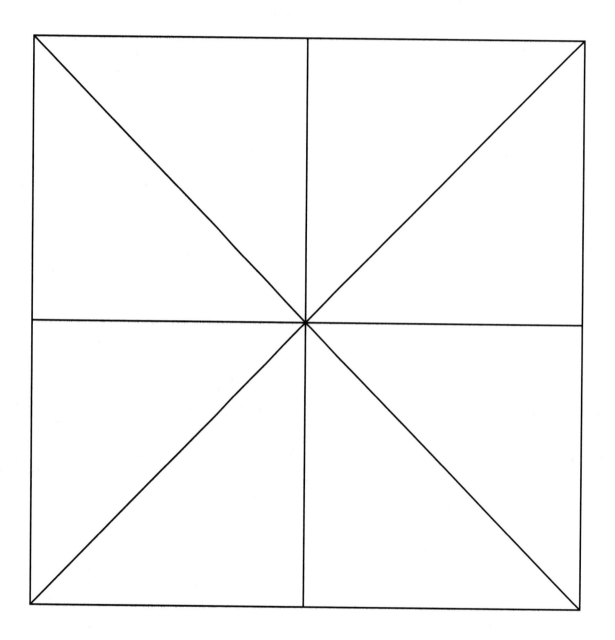

Tablita

Background

Tablita, a game from the Philippines, is played on a board of 26 squares. Although not seen frequently in market places, the game Culin (1900) reported is played throughout the country. Game implements included copper disks and a string or a strip of bamboo or rattan. When the betting aspect of tablita is removed for the classroom, an action game involving physical dexterity and estimation remains.

Materials

game board—wood, poster, or foam board (The game board may be drawn on the ground.)

string or ribbon

2 pennies

Number of Players: 2

Object

Each player attempts to toss a penny into a square on the board. Proximity to line segments forming the squares must be determined by estimation. Tablita may be played to an agreed-on sum.

How to Play

Position the string or ribbon about three feet, or 90 cm, above and across the middle of the board. This can be accomplished by placing the board on the floor between two desks and stretching the string or ribbon from desk to desk. The two players sit on the same side of the board. Each player tosses a penny over or as high as the string. If a player's penny lands on a line segment, no point is scored. If the penny lands within a square, a player scores one point. However, if the penny of each player lands within a square, the players must estimate which penny lands farther from the line segments. The player whose penny is farther from the line segments scores the point. Play continues to an agreed-on number of tosses to determine the winner.

Variation

Number the game board squares from 1 to 26 at random. At the end of the game, the winner is the player with the greatest total.

REFERENCE

Culin, Stewart. "Philippine Games." *American Anthropologist* 2 (October–December 1900): 643–56.

Tablita

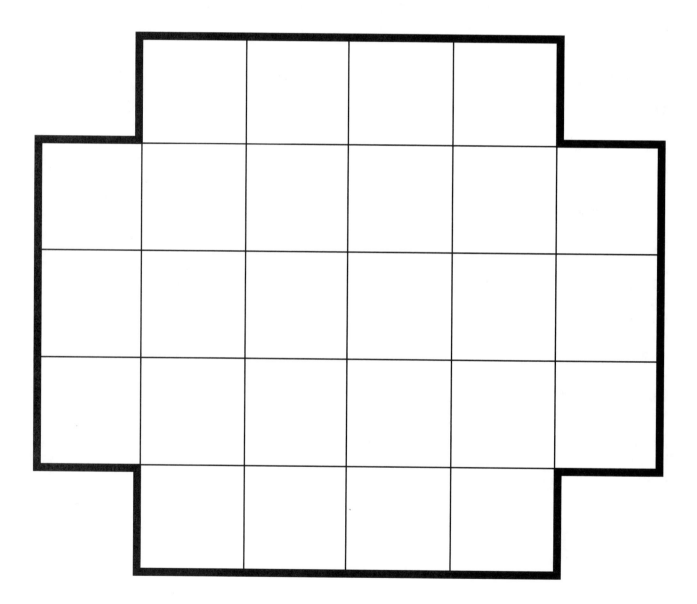

Taniko Patterns

Background

The Maori of New Zealand used flax fiber to weave decorative cloak borders called taniko. Triangles, zigzags, and diamonds form simple as well as complex designs. The patterns are in a straight line. This is the result of the way they are woven. In the eighteenth and nineteenth centuries, vegetable dyes were used to make red, orange, and yellow colors. Black fibers were obtained by placing the fibers in swamp mud. Today chemical dyes replace the old vegetable dyes.

Materials

Taniko Patterns activity page
colored pencils—red, orange, yellow, black
ruler

Object

Students will create their own straight-line taniko patterns.

Activity

Distribute the Taniko Patterns activity page. Discuss the sample patterns at the top of the activity page. Then have students create their own taniko patterns.

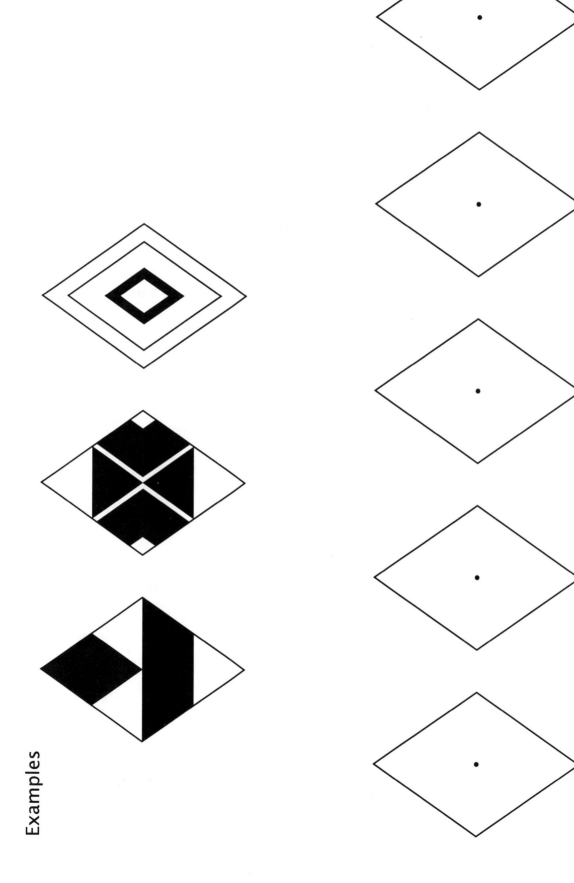

Examples

Create your own taniko patterns.

Mu Torere

Background

The Maori People live in New Zealand. It is believed that they migrated to New Zealand from Polynesian. According to Maori tradition, their ancestors came there in seven canoes. Bell (1979) notes that the Maori culture was "rich in action games and amusements." Mu Torere, however, seems to be the only game board the Maoris had. Game boards were marked on the ground, on the inner bark of the totara (a tall tree in New Zealand with hard, reddish wood), or with charcoal on hewn slabs of totara wood.

Materials

8-pointed star board
4 white playing pieces
4 black playing pieces

Number of Players: 2

Object

The player tries to block his opponent from moving. Playing strategies to win may be determined.

How to play

Position the white playing pieces on starpoints 1 through 4. Black playing pieces are on the remaining starpoints.

Alternate turns are taken. Black plays first. Jumps are not allowed.

Playing pieces may be moved

1. from one starpoint directly to an adjacent starpoint;
2. from the center of the board (putahi) to a starpoint;
3. from one of the starpoints to the center (putahi), if one or both of the adjacent starpoints are occupied by an opponent's playing piece or pieces

Sometimes the last rule (number 3 above) is kept only for each player's first two moves.

Let students discover that possession of the center (putahi) is necessary to win. Also players should attempt to get V shaped arrangements (e.g., the putahi and two circles at 45-degree angles) on the board.

If both players are aware of this and no mistakes are made, Mu Torere will be played to a draw.

REFERENCE

Bell, Robert C. *Board and Table Games from Many Civilizations*. Rev. ed. New York: Dover Publications, 1979.

Mu Torere

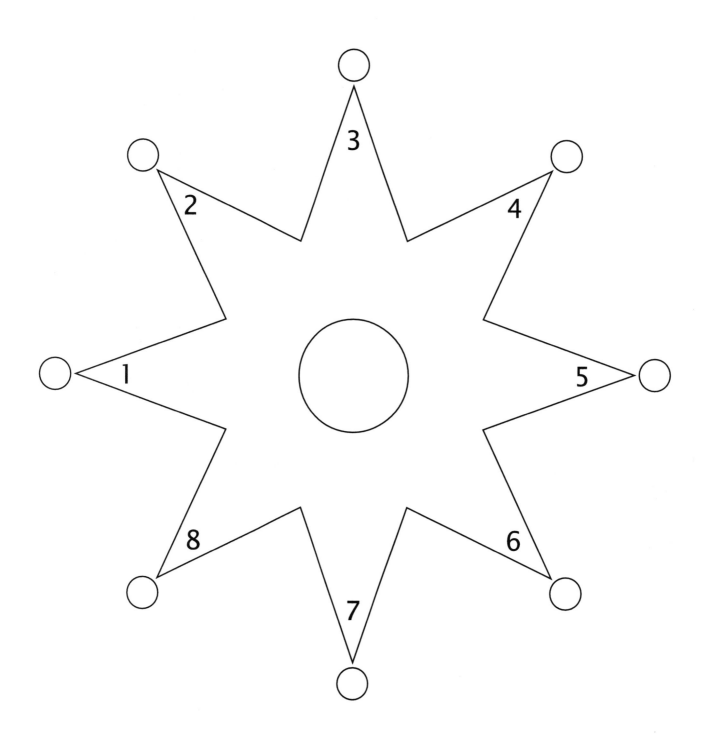

Nine Men's Morris

Background

Nine men's morris, also called *merels, mill,* and *morelles,* is an ancient game for two. This game can be traced back to about 1400 B.C. for the game board was one of the seven cut into the roofing slabs of the temple in Karnak, Egypt. Similar diagrams have been found in Sri Lanka, formerly Ceylon. There, two diagrams dating from A.D. 9–21 carved into the flight of steps on the lower part of the hill at Mihintale. Evidence shows that the game was known to the inhabitants of the first city of Troy as well as to the Vikings. A Bronze Age burial site in County Wicklow, Ireland, also revealed a morris board carved in stone.

According to Bell (1979), nine men's morris reached the zenith of its popularity in Europe during the fourteenth century. Even Shakespeare mentioned this game in 1598 in *A Midsummer Night's Dream.* In act 2, scene 1, lines 98–100, he wrote:

> The nine men's morris is fill'd up with mud,
> And the quaint mazes in the wanton green
> For lack of tread are undistinguishable.

Materials

nine men's morris board
9 red playing pieces
9 green playing pieces

Number of players: 2

Object

Nine men's morris is a reasoning and strategy game played on three concentric squares. Each player attempts either to block an opponent's playing pieces so that a move is not possible or to reduce the opponent to two playing pieces.

How to play

Players decide who plays first. The game begins with a clear board. Players take turns placing playing pieces, one at a time, at any angle or intersection of the figure. After all the playing pieces are on the board, play continues with pieces moved one at a time along a line to an adjacent empty position. Jumps are not allowed.

If a player succeeds in getting a mill—a row of three playing pieces on any line vertically or horizontally—he or she can remove one of the opponent's playing pieces from the board. However, if an opponent's playing piece is in a mill, it cannot be removed from the board.

New mills can be made by opening existing mills. That is, a playing piece can be moved to form a mill and then returned on the next turn. Each time a mill is made, the player is entitled to remove one of the opponent's playing pieces.

If a player has only three playing pieces left on the board and they form a mill, then on his or her turn the player must break the mill. This may result in the player losing the game.

A player wins when the opponent has only two playing pieces left on the board or is unable to move.

REFERENCES

Bell, Robert C. *Board and Table Games from Many Civilizations.* 1960. Reprint. New York: Dover Publications, 1979.

Shakespeare, William. *The Comedies of Shakespeare.* Introduction by A. C. Swinburne. London: Humphrey Milford, Oxford University Press, n.d.

Nine Men's Morris

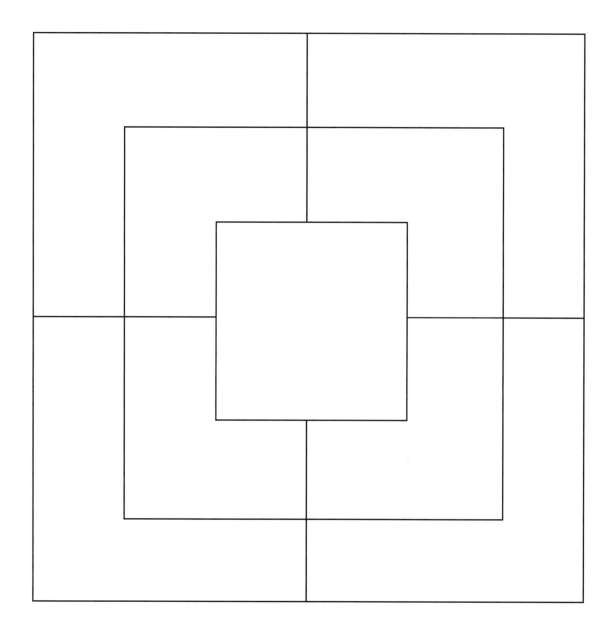

Julekurv

Background

Norway is a mountainous country extending into the Arctic Circle. Most of Norway borders Sweden, but in the north it borders Finland and Russia. The coastland of Norway is indented by many beautiful fjords—narrow inlets of the sea between high banks or cliffs.

The julekurv is a Norwegian Christmas basket. The basket is usually made from a flexible high-gloss paper that comes in many colors. Norwegians hang these baskets filled with sweet? on their Christmas trees.

Materials

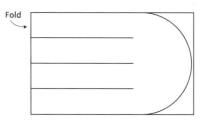
Fig. 1

2 strips of different colored construction paper 9 cm × 30 cm

construction-paper strip for handle

pencil

ruler having centimeter and millimeter divisions

scissors

glue

Object

Students measure in centimeters and use decimal fractions.

Activity

Cut 2 strips of construction paper 9 cm × 30 cm and perform the following steps on each strip:

1. Fold in half—9 cm × 15 cm.
2. Divide the strip into four equal parts along the fold. Each division will be approximately 2.2 cm.
3. Draw three 9.5-cm line segments extending from the three dividing points on the fold. See figure 1.
4. Cut along the three line segments.
5. Cut a semicircle opposite the fold.
6. To begin weaving the two strips X and Y together, position the pieces as in figure 2.
7. Place A inside 1, 2 inside A, A inside 3, and 4 inside A. One row is woven. Slide this row to the top of the three line segment cuts in strip X.
8. Place 1 inside B, B inside 2, 3 inside B, and B inside 4.
9. Place C inside 1, 2 inside C, C inside 3, and 4 inside C.
10. Place 1 inside D, D inside 2, 3 inside D, and D inside 4.
11. Attach handle.

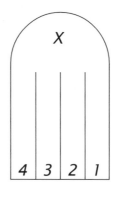

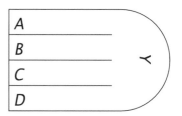

Fig. 2

51

Julekurv

Celtic Design

Background

The Celts dominated central Europe during the Iron Age. They migrated over Europe in the sixth and fifth centuries B.C. and reached the British Isles. Their movement spread the use of iron. King-like chiefs and priests known as Druids were part of their social structure. The Celts were well known for their beautiful ornamental art and colorful folklore.

Materials

Celtic ornamental design
Marking pen

Object

Students trace the Celtic ornamental design, a closed curve, without lifting their pencils.

Activity

Distribute the Celtic ornamental design. Discuss the two basic geometric notions involved in the design—point and curve. A point, an idea, may be thought of as a location in space. One point has one exact location in space. If the point is movable, then the path along which it moves is a curve. When one cannot tell where the path begins or ends, it is a closed curve. In other words, the beginning point and the endpoint of the path are the same. The Celtic ornamental design in the illustration is a picture of a closed curve, but it is not a simple closed curve. A simple closed curve does not cross over itself, and this Celtic design does cross over itself.

This type of Celtic design is common. An element of the design is the "knot." The interlacing may have been used as an aid in concentration—focusing the mind in a repetitive task.

Ask students if they think the design can be traced without lifting a marking pen. Tell them they may begin anywhere.

ADDITIONAL READING

Davis, Courtney. *Celtic Borders and Decoration.* London: Blandford, 1992.

Powell, Thomas G. E. *The Celts.* New York: F. A. Praeger, 1958. Reprint, New York: Thames & Hudson, 1997.

Celtic Design

Erin Go Bragh

Background

The national emblem of Ireland is the shamrock. The leaves of the shamrock plant have three leaflets. Legend tells us that Saint Patrick, who lived from about 385 to 461, used the shamrock to illustrate the doctrine of the Trinity. A shamrock, either real or artificial, is traditionally worn on 17 March, Saint Patrick's Day. Both Irish and non-Irish people can be seen wearing a shamrock on this day.

Materials

Erin Go Bragh page
ruler
green crayon or marking pen
scissors

Object

By connecting a given sequence of points with straight line segments, students construct shamrocks. The shamrock design gives the illusion of curves, although no curves are drawn.

Activity

Distribute materials. If desired, students can number the remaining sides that are not numbered. As the students work from the center of the design (9 to 1, 8 to 2, 7 to 3, and so on), discuss the acute angles that have sides in proportion. An equal number of marks are designated on both sides of the angles. After they have drawn the line segments, have students draw a stem. Students may color the shamrock, cut it out, and wear it on Saint Patrick's Day

Solution

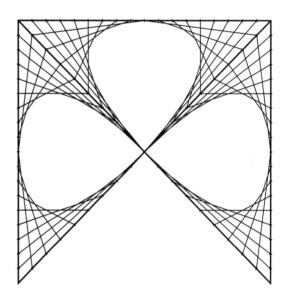

55

Erin Go Bragh

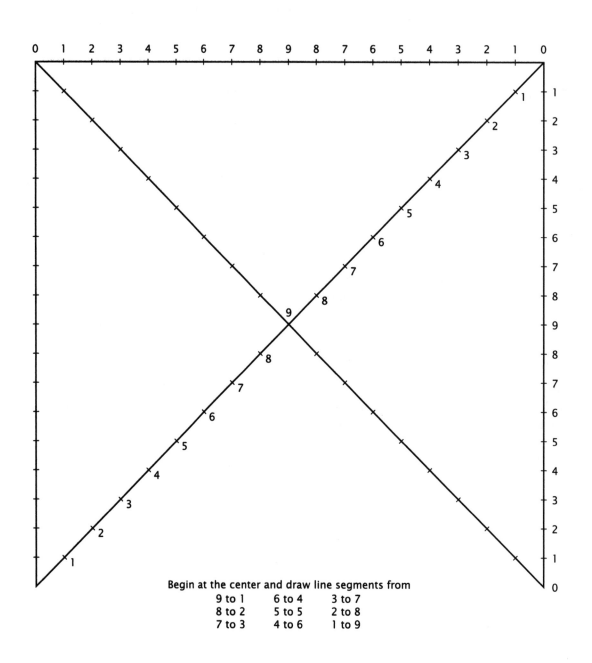

Begin at the center and draw line segments from

9 to 1	6 to 4	3 to 7
8 to 2	5 to 5	2 to 8
7 to 3	4 to 6	1 to 9

Roman Numerals

Background

Tradition says that Rome was founded by Romulus in 753 B.C. The Romans were ruled by the Etruscans until about 500 B.C., when the Romans overthrew the Etruscans and established a republic controlled by the patrician class. This republic survived for four centuries. The Roman Empire is dated from 27 B.C. and the rule of Augustus, the grandnephew of Julius Caesar, to the defeat and death in A.D. 476 of Romulus Augustulus, the last Emperor of the West.

Mathematics did not flower during the time of the Romans. This is surprising, since the Greeks had a great thirst for mathematical knowledge and the Romans often imitated the Greeks. According to Cajori (1980), the mathematics of the Romans probably came in part from the ancient Etruscans.

Their system is unusual in that the principle of subtraction is used. If a symbol of lesser value is placed before a symbol of greater value, then the lesser value is subtracted from the greater value. That is, XL means 50 – 10, or 40. But if the lesser value is placed after the greater value, the two values are added. That is, LX means 50 + 10, or 60. When a horizontal bar is placed over a symbol, the value is multiplied by 1000. The $\overline{M}$ means 1000 × 1000, or 1 000 000. The Roman system of numeration did not have a symbol for zero.

The Roman system of numeration is still in use today. Roman numerals are seen on clocks and watches and in book chapters, outlines, and formal inscriptions.

Roman Numerals

I	=	1
V	=	5
X	=	10
L	=	50
C	=	100
D	=	500
M	=	1000

Materials

Roman Numerals activity page

Object

The Roman Numerals activity page provides practice in renaming Hindu-Arabic numerals as Roman numerals. Also, one Roman numeral must be renamed as a Hindu-Arabic numeral. This activity can be used in conjunction with history, each date representing a famous historical event. (Other dates may be added or substituted as desired.)

Activity

Distribute the activity page and review with students the values of the seven Roman numerals given on the left. Review the principles of subtraction and addition used in the Roman system. Then ask students to rename each famous date as a Roman numeral.

REFERENCE

Cajori, Florian. *A History of Mathematics.* 3d ed. 1893. Reprint. New York: Chelsea Publishing Co., 1980.

Solution

MDCCLXXVI	— The signing of the Declaration of Independence
MCMLXIX	— Neil A. Armstrong and Edwin E. Aldrin, Jr., land on the moon.
MCMXXVII	— Charles Linbergh makes the first nonstop flight from New York to Paris.
MDCCLXXXIX	— George Washington inaugurated President
MMI	— The calendar year that the Royal Greenwich Observatory in Cambridge, England, officially adopted as the start of the third millennium
MDCCCIV	— Lewis and Clark expedition explores what is now the northwest United States.
MDCCCLX	— Pony Express between Sacramento, Calif., and St. Joseph, Mo., begins.

Roman Numerals

Fill in the blank spaces.

Roman Numeral	Value of Numeral	Date	Roman Numeral
I		1776	
V		1969	
X		1927	
L		1789	
C			MMI
D		1804	
M		1860	

Asalto

Background

Asolto is a variant of "fox and geese," a game from medieval times. Boards for "fox and geese" and asalto have 33 spaces. Bell (1983) shows an enlarged board game called asalto having 67 spaces. The enlarged board appeared during the mid-nineteenth century. Influenced by the Sepoy Rebellion against the British in India (1857–58), the game was renamed "Officers and Sepoys" (Grunfeld 1975). Basically both boards have a fortress defended by two or three "officers" against 24 or 50 "foot soldiers." Asalto is played in European countries (e.g., Germany, France, and England).

Materials

game board
2 officer playing pieces (discs or penny tiles)
24 foot soldier playing pieces (a different color)

Number of Players: 2

Object

Asalto is a game of logic where planned movements in space determine the winner.

How to Play

To begin, one player—representing the officers—places the two officers any place within the 3 × 3, star-like fortress. The other player—representing the foot soldiers—places the 24 foot soldiers on the remaining spaces outside the fortress.

Moves are made one space at a time. Officers may move in *any* direction; foot soldiers may move only toward the fortress in a straight line or diagonally. Officers may capture a foot soldier by jumping over the foot soldier to an empty space. A captured foot soldier is removed from the board. If an officer fails to see a possible capture, the officer is removed from the board.

Foot soldiers can win by (1) trapping the officers so movement is impossible or (2) occupying every spot within the fortress. Officers win when the foot soldiers have become so few that it is impossible for them to win.

Comment to students: Asalto is a game where one side greatly outnumbers the other. Ask them to consider if it is a given that the foot soldiers will win. Discuss the reasons for their answers.

REFERENCES

Bell, Robert C. *The Boardgame Book.* New York: Exeter Books, 1983.

Grunfeld, Frederic V., ed. *Games of the World.* New York: Holt, Rinehart & Winston, 1975.

Asalto

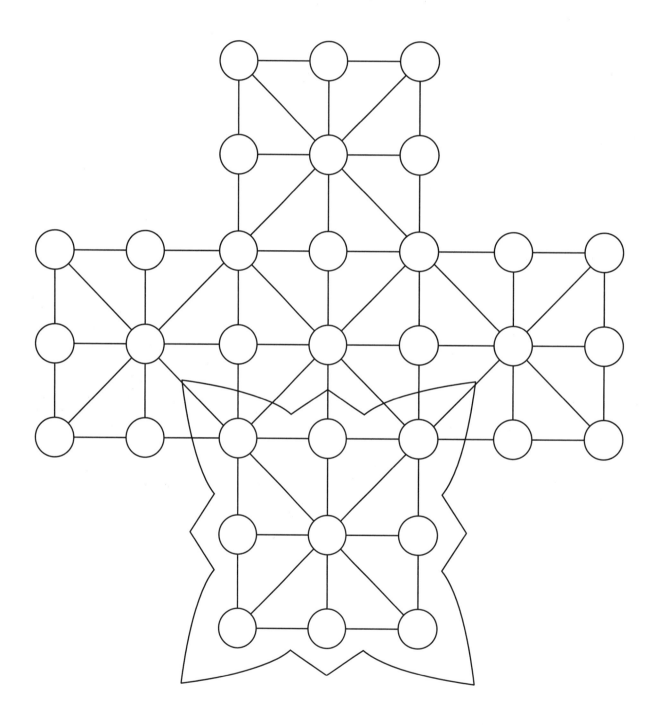

Dreidel

Background

A dreidel is a four-sided top. Hebrew letters are on the four sides of the dreidel:

 nun   *gimmel*   *hay*   *shin*

These symbols are the initial letters of the Hebrew words *nes gadol hayah sham* "a great miracle happened there," referring to an event that took place over two thousand years ago. After Judas Maccabee and only a few soldiers defeated the army of a cruel king who attempted to make the Jewish people worship idols, the Jews lighted the temple lamp. There was only oil in the lamp for a few hours and yet the lamp burned for eight days and eight nights. This is why the Festival of Lights, Hanukkah, is celebrated for eight days.

Materials

1 dreidel (inexpensive dreidels can be purchased—contact a local synagogue for information.)

10 pieces of "Hanukkah gelt" for each player (Dried beans, peanuts, candy, small stones, or pennies can be used.)

Number of players: 2 or more

Object

The dreidel game is a game of chance providing practice in adding, subtracting, and using the common fraction 1/2.

How to play

Before beginning play, each player places an agreed-on number of beans in the center of the table or in a pot. A turn consists of a player spinning the dreidel and follow these directions:

1. If *nun* turns up, nothing is taken from the pot. The next player then spins.

2. If *gimmel* turns up, the player takes all the beans in the pot. When the pot is empty, each player again places the earlier specified amount in the pot. The next player spins.

3. If *hay* turns up, the player takes half the beans in the pot. For an odd number, the player takes half plus one. The next player then spins.

4. If *Shin* turns up, the player adds the agreed-on number of beans to the pot. The next player then spins.

The game is over either when one player has won all the beans or after a designated number of rounds. One round is completed after each player has spun the dreidel once.

Variation

Give each letter a numerical value and play to get the highest score, such as 1000. Or the winner can be the player with the highest score after a designated number of rounds.

nun	50	*hay*	3
gimmel	5	*shin*	300

Letters		Expected Outcomes
ב	(nun)	
ג	(gimmel)	
ה	(hay)	
ש	(shin)	
	TOTAL	100

After playing

Have older students experiment as follows:

1. If you spin a dreidel 100 times, how many times would you expect the dreidel to turn up, *nun, gimmel, hay,* and *shin?* Record what you expect.

Note that the outcomes are equally likely.

2. Try this experiment and see what happens. Compare your results with what you expected. Tell what happened.

3. Play the game again. Keep a record of each player's spins. Do the results surprise you?

ADDITIONAL READING

Goodman, Philip. *The Hanukkah Anthology.* Philadelphia: Jewish Publication Society of America, 1976.

Dreidel Decoration

Materials

construction-paper rectangles (sizes of rectangles can vary)

ruler

scissors for each student

tape

Object

Students will make a paper dreidel decoration from a construction-paper rectangle.

Activity

1. Tell students to estimate and draw $\overline{FE}$ a little less than one third from the bottom of the rectangle.
2. Divide the smaller rectangle $CDFE$ in two parts (see $\overline{HG}$.)
3. Draw diagonals from F to G and E to G.
4. Cut on the diagonals. The congruent triangles (see 1 and 2) are placed at the top with tape.
5. The holiday dreidel decoration is ready for display.

Dreidel Decoration

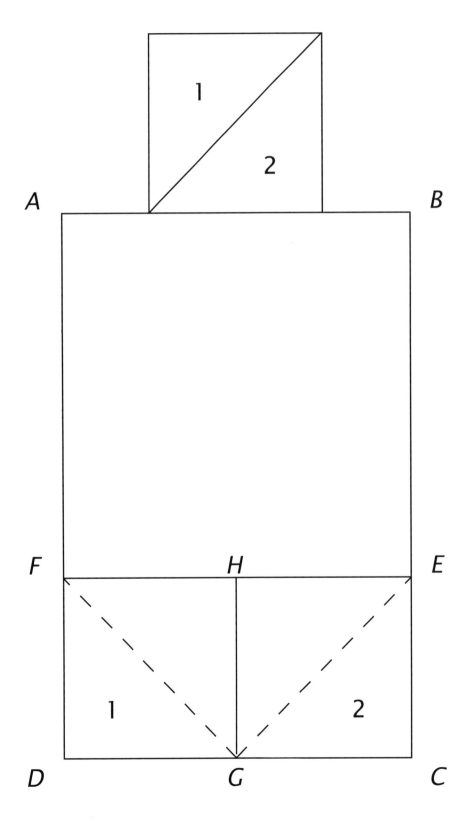

Magen David

Background

The Magen David, a six-pointed star, is also called the Star of David or the Shield of David. It is a symbol of Judaism, and one appears on the Israeli national flag.

Object

Using their listening skills, students will follow directions to make a six-pointed star on a clock face.

Materials

Magen David activity page

ruler

pencil or marker

Activity

Say the following instructions to students:

1. Label the circle like the face of a clock. Twelve o'clock is done for you.
2. Begin at 12 o'clock and continue clockwise.
3. Draw a line segment from 12 o'clock to the time that is 4 hours later. (4 o'clock)
4. Draw a line segment from this time to the time that is 4 hours later. (8 o'clock)
5. Draw a line segment from this time to the time that is 4 hours later.
6. What time is it now? (12 o'clock)
7. What shape is drawn on the clock face? (equilateral triangle)
8. Begin at 6 o'clock and continue clockwise again.
9. Draw a line segment from 6 o'clock to the time that is 4 hours later. (10 o'clock)
10. Draw a line segment from this time to the time that is 4 hours later. (2 o'clock)
11. Draw a line segment from this time to the time that is 4 hours later.
12. What time is it now? (6 o'clock)
13. What is the second shape you have just drawn on the clock face? (equilateral triangle)
14. What do these two equilateral triangles form? (six-pointed star)

Magen David

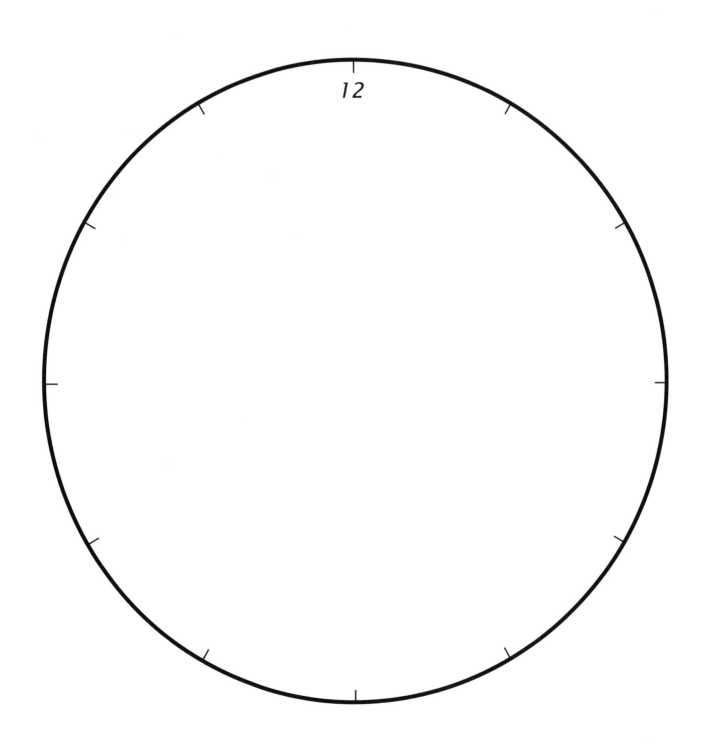

Solution

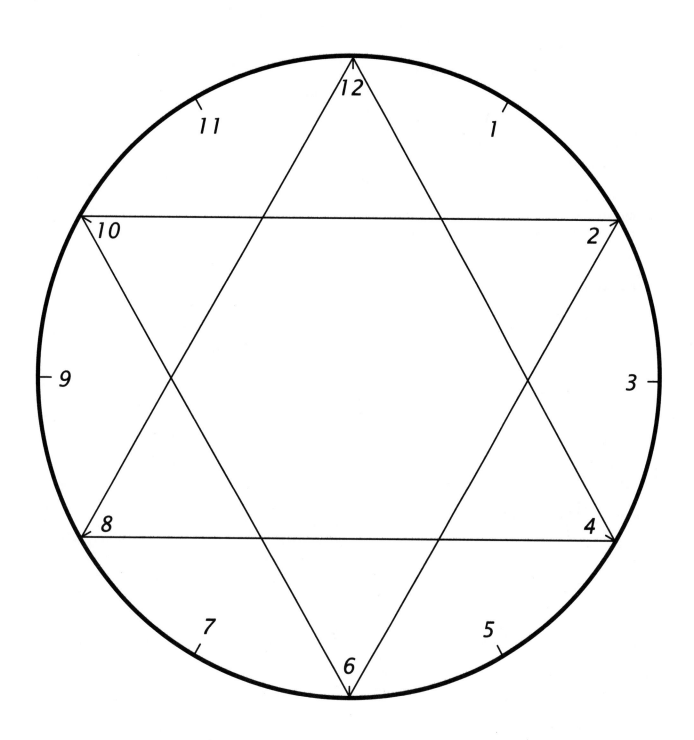

Tessellations

Background

Our word *tessellate* comes from the Latin *tessella,* diminutive of *tessera,* "a small square block." The noun *tessellation* comes to us from the Latin *tessellare,* "to pave with tiles."

Tessellations have been used for many centuries. The Romans used small tiles to make mosaics for their floors. In 1941, excavations on the south side of the Cologne cathedral in Germany revealed the remains of a second-century Roman house and its now famous floor, known as the mosaic of Dionysos. The mosaic measures 74 meters square. This floor mosaic is composed of approximately two million small blocks of limestone, marble, slate, pottery, and blue or green glass pastes.

Muslim artisans have demonstrated exceptional skill in juxtaposing shapes in coherent geometrical patterns. The Islamic religion forbids the use of the human form in art. Arabic buildings are often decorated with colorfully tiled facades. The interiors may also contain exquisite decorative mosaics and carvings.

When students are studying tessellations, the works of the Dutch artist M. C. Escher (1898–1972) may be of interest. Escher created fascinating pattern designs.

Materials

construction-paper shapes, such as squares, triangles, rectangles, and hexagons

tessellation activity pages

tessellation background grid

ruler

compass

colored pencils or markers (optional)

Object

Students will complete pattern designs—tessellations. Students should also be encouraged to develop a pattern design of their own on the tessellation background grid.

Activity

Study the geometric shapes that fit together without any spaces or gaps. Students can demonstrate with the construction-paper shapes which shapes will fit together. Discuss the background and meaning of the word *tessellation.* Ask students if they have noticed in their surroundings any pattern designs composed of geometric shapes. Flooring, chicken wire, and checkerboards are a few examples.

Distribute a tessellation activity page and consider with the students the shape formed in relationship to the square. Interestingly, a more exotic design often has as its basis a simple shape. Have students complete the design.

Equilateral triangles, squares, and regular hexagons form tessellations. A regular polygon forms a tessellation if the measure of its vertex angle is a divisor of 360. For example, the measure of the vertex angle of an equilateral triangle is 60 degrees; of a square, 90 degrees; and of a hexagon, 120 degrees. Sixty, 90, and 120 are all divisors of 360. The measure of the vertex angle of a regular pentagon, however, is 108 degrees. One hundred eight is not a divisor of 360. A tessellation using only regular pentagons cannot be formed. Two or more different regular polygons having sides the same length may be used to create other beautiful tessellations. Distribute an activity page on which two different shapes fit together for a more complex design.

ADDITIONAL READING

Mold, Josephine. *Tessellations.* Cambridge, U. K.: Cambridge University Press, 1969.

Tessellation Activity Page

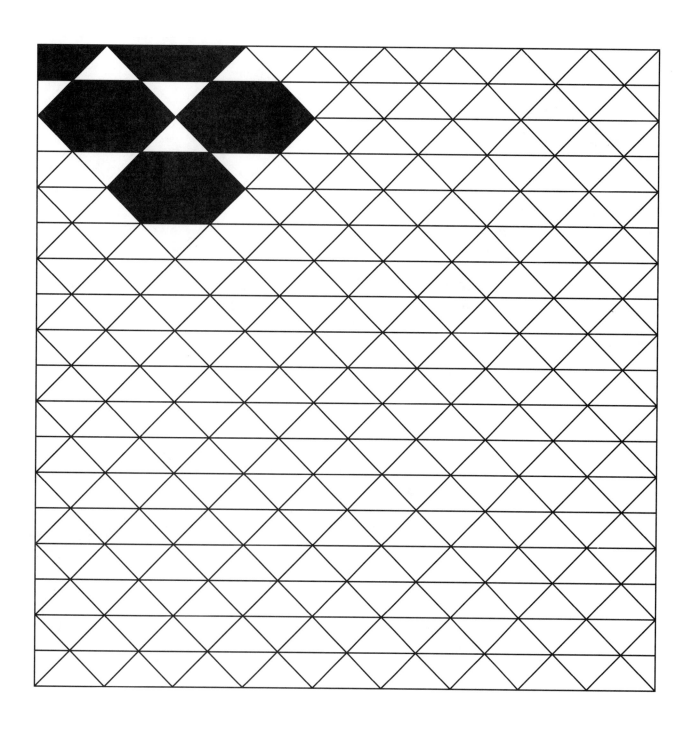

Tessellation Activity Page

Tessellation Background Grid

Wind Rose

Background

Although it is not known who invented the magnetic compass or when, many claim to be the original inventor (Greenhood 1964). The Arabs, the Chinese, the Etruscans, the Finns, the Greeks, and the Italians all claim this distinction. The diagram that appears on charts and compass cards is called a "wind rose" or "rose of the winds" because of its flowerlike appearance. Eight principal winds—north, south, east, west, northeast, southeast, southwest, and northwest—are charted. An additional eight points, the half-winds, bring the number of points to sixteen. Sixteen more points designate the quarter-winds for a total of thirty-two. Using only a compass and ruler, students may construct a wind rose that has all thirty-two points (Krause 1973). Historically, wind roses were colored vividly. The principal winds were colored gold or black; the half-winds, green or blue; and the quarter-winds, red.

Materials

Wind Rose activity page

directions sheet

ruler

colored pencils markers, if desired

Object

Students can construct a wind rose of sixteen points by following a pattern and drawing line segments from point to point.

Activity

Distribute the Wind Rose activity page and the directions sheet. Ask students to fill in the blanks on the direction sheet. Discuss the patterns they see. Have them complete the Wind Rose activity page, label the eight principal winds and the eight half-winds, and color the wind rose if desired. Ask students to consider why this diagram is called a wind rose.

REFERENCES

Greenhood, David. *Mapping.* Chicago: University of Chicago Press, 1964.
Krause, Marina C. "Wind Rose, the Beautiful Circle." *Arithmetic Teacher* 20 (May 1973): 375–79.

Directions

Draw line segments from:	Sum	Difference
1 to 8	9	7
2 to 9		
3 to 10		
4 to 11		
5 to		
to 13		
7 to		
8 to		
to		

Draw line segments from:	Sum	Difference
10 to 1		9
to 2		
12 to		
to 4		
to 5		
15 to		
to		

Wind Rose

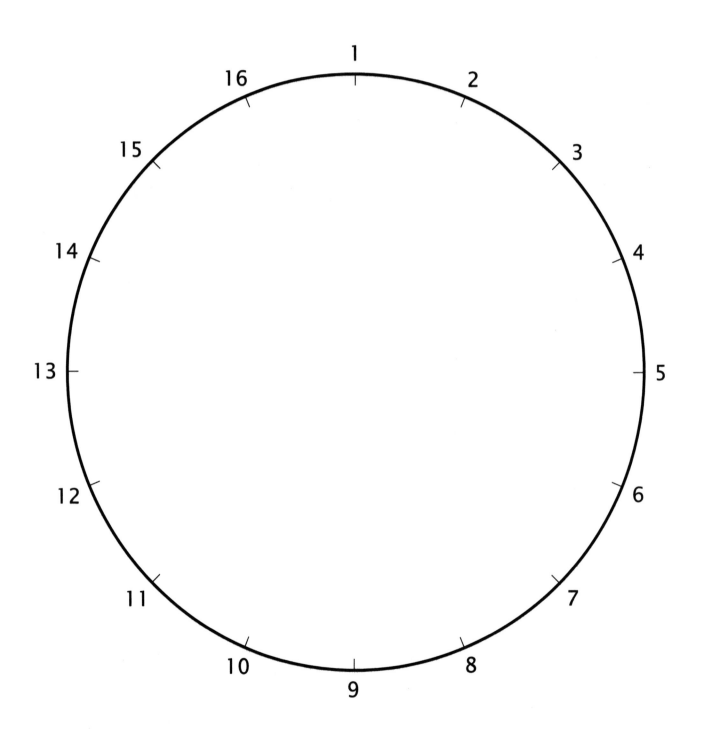

Wind Rose Solution

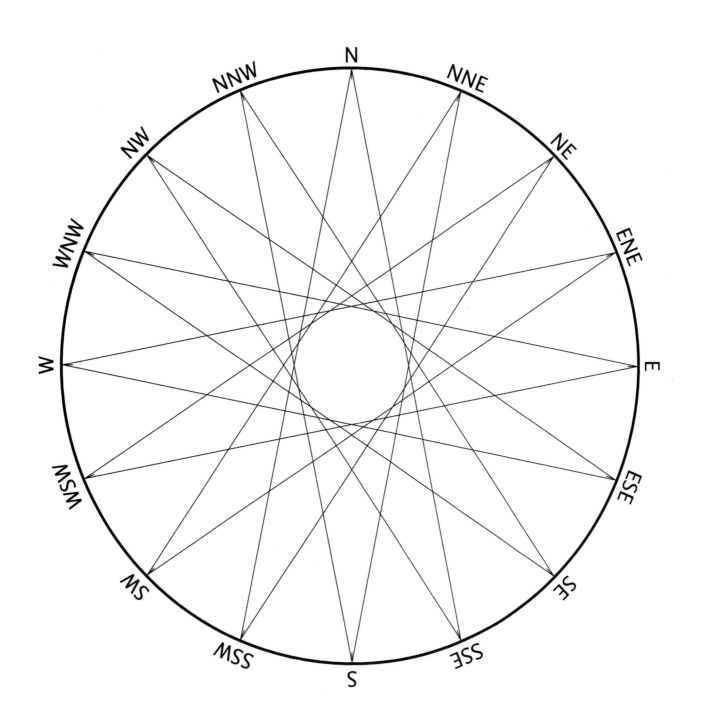

Quipu

Background

The Incas used *quipus,* pronounced kē' pōos, for keeping records and sending messages. Information such as population and resources was recorded on the quipu. The quipu was a counting device based on the decimal system. Because the Incas left no written record, the method of deciphering is not fully known.

The quipu consisted of a top cord or rope to which smaller cords or strings of different colors were attached. Authorities believe that a string's color, position, thickness, and the number of knots tied in it were all significant.

According to Bernand (1994), some of the strings were a single color; others were two, three, or more colors combined. The color referred to the item. For example, gold was represented by a yellow string, silver by a white string, and soldiers by a red string. Items having no color were positioned by their importance, from the most important first to the least important. The number of ones, tens, hundreds, thousands, and tens of thousands were depicted by the knots. (fig. 1). Higher numbers were rarely used although the Incas were able to express greater numbers.

Materials

1 piece of rope or cord

strings or yarn of different colors (Use one string for each place value being studied.)

Object

Students represent numbers in base ten.

Activity

Have the students make a quipu for base ten. The number of strings attached will depend on the place values the students are studying. Assign place values to the different colored strings. Tie knots in the strings to represent different numbers.

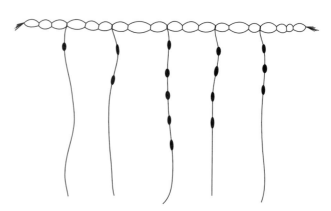

Fig. 1 A quipu representing 12 543.

REFERENCE

Bernand, Carmen. *The Incas: People of the Sun.* New York: Harry N. Abrams, 1994.

Inca Bird Design

Background

The center of Inca empire was Cuzco, Peru. The Incas founded Cuzco during the eleventh century. Their empire extended over 3000 kilometers, north to south, along the western coast of South America. The land of the Incas also reached inland and included the Andes Mountains.

The word *Inca* may refer to the emperor, who ruled by divine right, but it is commonly used by historians to refer to the people or the empire itself. The Inca empire of six million people was conquered in 1533 by Francisco Pizarro and his force of fewer than two hundred Spanish soldiers.

The Incas are well known for their network of roads, building constructions, pottery, weaving, and uses of copper, bronze, tin, silver and gold. The Inca bird of this activity was a woven design.

Materials

Inca bird design
transfer grid

Object

Using a grid system of squares, the student transfers and enlarges the design. If placed on the large grid, the design could be *reduced* by transferring it to the small grid.

The mathematical basis for scale drawings is similarity. The ratio of similarity is the scale given for a drawing.

The small grid and the large grid are similar figures; that is, they have the same shape. The scale is doubled, since the length of the side of the large square is double the length of the side of the small square. But the area of the large shape is four times that of the small shape. Note that four 1 cm × 1 cm shapes will fit on one 2 cm × 2 cm shape.

Activity

Distribute the Inca bird design and the transfer grid. Discuss the uses of scale in our lives—for example, diagrams, maps, or photographs that are either enlarged or reduced. Through the centuries, travelers have reached their destinations using scale to determine distance and direction. Scale enables the traveler to look at a map and measure the distance between points. Perhaps the most familiar kind of scale is the "inches-to-the-mile" scale. It is an amazing fact that one inch of paper can represent one mile, or 63 360 inches, of our planet. Also, one centimeter of paper can represent ten kilometers, or 1 000 000 centimeters, of our planet.

After the students have enlarged the design, ask about the scale and area of the new shape.

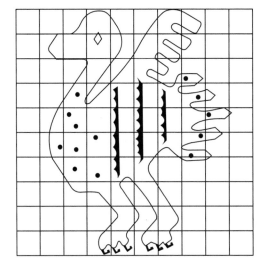

Enlarge the design by drawing it on the transfer grid. Use the following key: 1 small unit → 1 large unit.

ADDITIONAL READING

Glubok, Shirley. *The Art of Ancient Peru.* New York: Harper & Row, Publishers, 1966.

Additional Activity Page

Background

This stylized duck is based on an ancient Inca design.

Materials

Inca Duck activity page

Object

This activity page provides practice in looking for geometrical elements such as shapes and line segments within a design.

Activity

Distribute the activity page and ask students to study the design and tell what geometrical shapes and line segments they see. They may see concentric circles (head), triangles (head and tail), an oval (body), parallelograms (wing and neck), rectangles (tail), and parallel line segments.

Inca Duck

Colombian Star

Background

The Republic of Colombia is located in the northwest corner of South America. It is the only country on the South American continent that fronts two oceans—the Pacific and the Atlantic (Caribbean Sea). Bogotá, the capital, at an altitude of 8560 feet (2610 meters), has been called the "Athens of South America" because it has been a cultural center from the time of the Spanish viceroys.

The Colombian Star was shared with the author by a fourth-grade boy who had recently moved to the United States from Colombia. Holding the star he had just made, the boy said, "This is what we do in Colombia. We put them up in our windows at Christmas."

Materials

paper squares, any size (The size of the star can vary.)

Object

Students will construct a four-pointed Colombian Star. The folding activity is geometrical and involves estimating the fraction 3/4. This activity also provides practice in following oral directions.

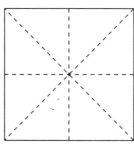

Fig. 1.

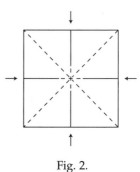

Fig. 2.

Activity

1. Fold the square horizontally, vertically, and diagonally (fig. 1).
2. Cut along the horizontal and vertical folds from the edge of the square to about 3/4 of the distance to the center (fig. 2).
3. Make an airplane fold on each of the ends of each diagonal (fig. 3). Four airplane folds will be made.
4. The Colombian Star is now ready to display.

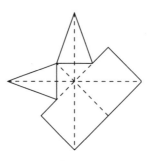

Fig. 3.

Fig. 4.

The Maya

Background

The ancient Mayan culture flourished in Central America and southern Mexico between approximately A.D. 250 and 900. The Mayan numeration system included the use of zero, place value, addition, and the repetition of symbols. The system was essentially vigesimal, or base twenty. Figure 1 shows an exception to the vigesimal system: 18 *uinals,* not 20, make 1 *tun.*

The Maya were fascinated by the passage of time and often expressed time intervals. They had two sophisticated calendars: the *haab,* a 365-day year with fifty-two-year cycles, and the *tzolkin,* a sacred or ceremonial year of 260 days. The tzolkin had thirteen 20-day months. The corrected calendar year with eighteen 20-day months plus one 5-day month approximates the astronomically determined year with greater accuracy than the Gregorian calendar does.

The Maya wrote their numerals vertically. That is, reading from bottom to top the values are as follows: 1's, 20's, 360's, 7 200's, and so on. Three symbols were used in counting: a dot represented 1, a horizontal bar represented 5, and a shell represented 0. (See figs. 2 and 3.)

20 kins, or days	= 1 uinal, or 20 days
18 uinals	= 1 tun, or 360 days
20 tuns	= 1 katun, or 7 200 days
20 katun	= 1 cycle, or 144 000 days
20 cycles	= 1 great cycle, or 2 880 000 days

Fig. 1.

Materials

Maya Numerical Symbols sheet
calendar grid

Object

Students are given an opportunity to work with a numeration system other than base ten by constructing a calendar using the Mayan numerical symbols. The Mayan numeration system, essentially a base of twenty, had the position of the values written vertically.

$$
\begin{aligned}
1 \times 360 &= 360 \\
3 \times 20 &= 60 \\
7 \times 1 &= \underline{7} \\
&\ 427
\end{aligned}
$$

Fig. 2.

Activity

Discuss the Maya and their numeration system. Distribute the Maya Numerical Symbols sheet and point out the placement of the values vertically. Twenty, for example, is written as one 20 and zero 1's. One hundred is written as five 20's and zero 1's. Have students practice writing the Mayan symbols for 1 through 31.

Distribute the calendar grid. Have the students select a month and create a Mayan calendar of their own using the Mayan numerical symbols.

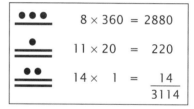

$$
\begin{aligned}
8 \times 360 &= 2880 \\
11 \times 20 &= 220 \\
14 \times 1 &= \underline{14} \\
&\ 3114
\end{aligned}
$$

Fig. 3.

ADDITIONAL READING

Cajori, Florian. *Notations in Elementary Mathematics.* A History of Mathematical Notations, vol. 1. LaSalle, Ill.: Open Court Publishing Co., 1928.

Smith, David Eugene. *Special Topics of Elementary Mathematics.* History of Mathematics, vol. 2, 1925. Reprint. New York: Dover Publications, 1958.

Maya Numerical Symbols

Hindu-Arabic	Maya
0	👁
1	•
2	• •
3	• • •
4	• • • •
5	___
6	• ___

Hindu-Arabic	Maya
7	• • ___
8	• • • ___
9	• • • • ___
10	≡
15	≡
20	• 👁
100	👁

My Maya Calendar For _____

SUNDAY	MONDAY	TUESDAY	WEDNESDAY	THURSDAY	FRIDAY	SATURDAY

PATOLLI

Background

The Aztecs were Indian people who dominated central Mexico when the Spaniards under Cortés arrived in 1519. The Aztecs played patolli on a mat game board in the shape of a cross. Within the cross, line segments formed squares, or houses. The patolli game board resembles the *pachisi* board. Beans called *patolli* were used as one-sided dice.

The earliest Spanish account of patolli was written by Francisco Lopez de Gomara, a Spanish chronicler, in 1552, thirty-one years after the conquest, and reveals that Emperor Montezuma sometimes watched as the game was played.

Before the game, players would bring fire, toss incense into the flames, and make offerings of food to Macuilxochitl ("Five Flowers"), the god of the dice. During the game, players would call on not only Macuilxochitl but also Ometochtli, the god of gambling, to give them good points on the dice.

The game was played enthusiastically, and betting was heavy. After the conquest, the Aztecs stopped playing the game, apparently because of their superstitions.

Since too few of the original rules of the game are known for play, the rules below are suggested. Players can add penalties to the game if desired.

Materials

game board

5 lima beans

Mark a single dot on one side of each bean with a marking pen.

If there are two players, each needs six playing pieces of the same color; if three players, four playing pieces each; and if four players, three pieces each. Playing pieces could be small colored stones or construction-paper shapes.

Number of players: 2 to 4

Object

Patolli is a game of chance in which moves are determined by the toss of five beans.

How to play

Each player has his or her own starting place at a different arm of the game board A, B, C, or

D. Players take turns. For the toss, the beans are held in both hands. Determine the order of play by tossing the beans and counting the dots facing up. If players tie, they toss the beans again. The highest scorer plays first and tosses the beans again for the first turn.

The first playing piece can be brought onto the game board with any toss. The remaining pieces can be brought onto the game board only when a 1 is tossed. Count the tosses as follows:

    1 dot up, move forward 1 space
    2 dots up, move forward 2 spaces
    3 dots up, move forward 3 spaces
    4 dots up, move forward 4 spaces
    5 dots up, move forward 10 spaces

The first player to move all his or her playing pieces completely around the patolli game board is the winner. To leave the board, the player must have the exact toss for the number of spaces left.

ADDITIONAL READING

Tylor, E. B. "On the Game of Patolli in Ancient Mexico, and Its Probably Asiatic Origin." *Journal of the Anthropological Institute of Great Britain and Ireland* 8 (1879): 116–31.

Palolli

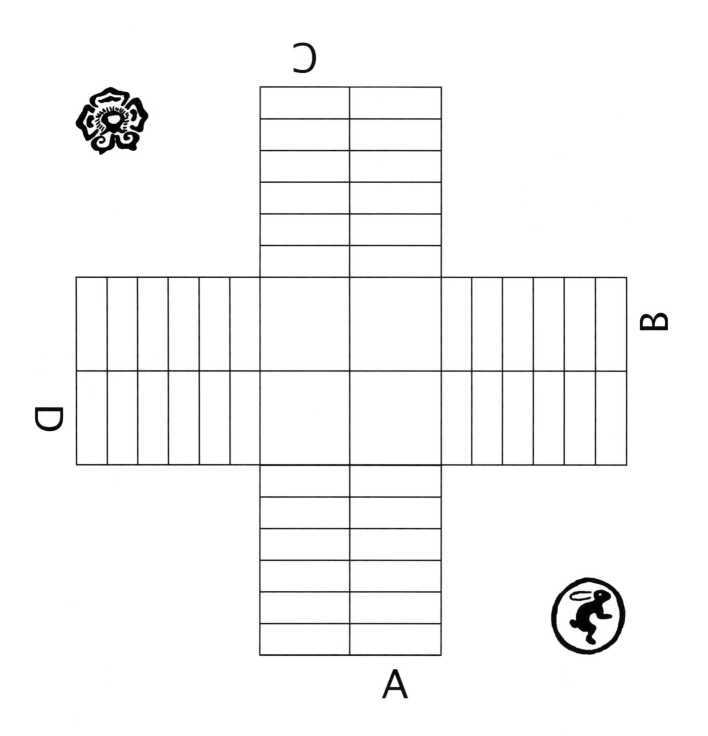

Aztec Calendar

Background

The Aztecs lived in the Valley of Mexico and surrounding areas from about A.D. 1200 until 1521, when they were conquered by the Spanish under the leadership of Hernando Cortés. (From this conquest came the beginnings of the Mestizo people and the Mexico of today.) The capital city of the Aztec empire was Tenochtitlán, pronounced tay-nohch-TEE-tlahn, which is now Mexico City.

One of the most magnificent expressions of Aztec art is the Sun Stone, or Aztec calendar (fig. 1). The calendar attests to the Aztecs' knowledge of astronomy and mathematics. Hieroglyphs for their days, months, and suns (or cosmic cycles) are pictured on the calendar.

The carved stone is 3.7 meters (about 12 feet) in diameter and has a mass of twenty-four metric tons. It took fifty-two years to make the calendar, from 1427 to 1479. Most authorities believe only stone tools were used. This calendar is superior to calendars used by Julius Caesar or Cleopatra and is 103 years older than the Gregorian calendar created by Pope Gregory XIII in sixteenth-century Rome. The Gregorian calendar is used worldwide today.

Originally the Aztec calendar was placed atop the main temple in Tenochtitlán. Today, Mexico City's cathedral stands on the site. The Aztec calendar faced south in a vertical position and was painted in brilliant red, blue, yellow, and white.

Tenochtitlán fell in 1521 to the Spaniards, who destroyed the temple and buried the stone. The stone was lost for more than 250 years; in December 1790 it was found by accident during repair work on the cathedral. Today, the Aztec calendar is in the National Museum of Anthropology in Mexico City.

The face of Tonatiuh, the Aztec sun god, is in the center circle of the stone. Around the face are four squares called *nahui-ollin*, or "four movement." According to Aztec legend, these squares represented the different ways that the four previous suns (or worlds) had come to an end: first by wild animals, then by wind, by fire, and by floods. The Aztecs believed they were living in the fifth and last world. They believed the fifth world would be destroyed by earthquake.

Continuing outward, the next concentric circle shows twenty squares, each naming one of the twenty different days of the Aztec month. Clockwise these days are as follows:

1. Snake—Coatl
2. Lizard—Cuetzpallin
3. House—Calli
4. Wind—Ehecatl
5. Crocodile—Cipactli
6. Flower—Xochitl
7. Rain—Quiahuitl
8. Flint—Tecpatl
9. Movement—Ollin
10. Vulture—Cozcacuauhtli
11. Eagle—Cuauhtli
12. Jaguar—Ocelotl
13. Cane—Acatl
14. Herb—Malinalli
15. Monkey—Ozomatli
16. Hairless Dog—Itzquintli
17. Water—Atl
18. Rabbit—Tochtli
19. Deer—Mazatl
20. Skull—Miquiztli

Aztec Calendar (Continued)

The Aztec year consisted of eighteen months, each having 20 days. Each month was given a specific name. This arrangement took care of 360 days (18 × 20), to which five dots were added inside the circle. These dots, known as *nemontemi*, were sacrificial days.

The next concentric circle is composed of square sections with five dots in each section, probably representing weeks of five days. Next are eight angles dividing the stone in eight parts. These represent the sun's rays placed according to the cardinal points.

On the lower portion of the stone, two enormous snakes encircle the stone and face each other. Their bodies are divided into sections containing the symbols for flames, elephant-like trunks, and jaguar-like forelegs. It is believed that these sections are also records of fifty-two year cycles. A Square is carved at the top of the calendar between the tails of the snakes. Inside the square the date 13 *acatl* (or "cane") is carved. This corresponds to 1479, the year the calendar was finished.

Eight equally spaced holes appear on the very edge of the calendar. The Aztecs placed horizontal sticks here, and the shadows of the sticks would fall on the figures of the calendar; thus the stone also served as a sundial.

Materials

Aztec Numerical Symbols sheet
calendar grid

Object

Students will construct a calendar using Aztec numerical symbols. The Aztec numeration system had a base of twenty. Units of one were represented by a dot. A diamond stood for 10; a flag, 20; a feather, 400; and a bag with tassels, 8000. The Aztecs applied repetitive and additive principles: for example, 423 is 400 + 20 + 3, or a feather, a flag, and three dots.

Activity

Discuss the Aztecs and their calendar. Distribute the Aztec Numerical Symbols sheet and discuss the repetitive and additive principles of the numeration system. Have students practice writing the symbols 1–31 in the Aztec system.

Distribute the calendar grid. Have the students select a month and create an Aztec calendar of their own using the Aztec numerical symbols; check that students are beginning the calendar with the first day in the correct posi-

tion on the grid. The name of the month should be written at the top of the calendar. Students may also decorate the top of the calendar with flowers—the Aztecs loved flowers.

ADDITIONAL READING

Dávila, Francisco González. *Ancient Cultures of Mexico*. Mexico City: Museo Nacional, 1961.

Pasztory, Esther. *Aztec Art*. New York: Harry N. Abrams, 1983.

Ragghianti, Carlo Ludovico, and Licia Ragghianti Collobi. *National Museum of Anthropology Mexico City*. Milan: Newseek & Arnoldo Mondadori Editore, 1970.

"Stone of the Fifth Sun." *National Geographic* 158 (December 1980): 757–59.

Fig. 1.

Aztec Numerical Symbols

Hindu-Arabic	Aztec
1	•
2	••
3	•••
4	•• ••
5	•• •••
6	••• •••
7	••• ••••
8	•••• ••••

Hindu-Arabic	Aztec
9	⦂⦂⦂
10	◇
20	⚐
100	
200	
300	
400	
8000	

My Aztec Calendar For _____

SUNDAY	MONDAY	TUESDAY	WEDNESDAY	THURSDAY	FRIDAY	SATURDAY

Toma Todo

Background

Mexico, the land of the ancient Aztecs, lies between the United States and Central America. It is a land of strikingly beautiful contrasts—snow-capped mountains, volcanoes, deserts, and beaches. Mexico City, the capital, has an elevation of about 7800 feet (2380 meters) with spring-like climate. "Toma todo" or "pirinola" is a game of chance from Mexico. Children as well as adults play Toma todo with a six-sided spinner (pirinola) or top. Adults often play the game using money.

Materials

some beans or wrapped candies (suggested number: 10 per player)

1 six-sided pirinola (Inexpensive pirinolas can be purchased in shops or areas that sell items made in Mexico. For example, shops along Olvera Street in Los Angeles sell various kinds.)

Object

Toma Todo is a game of chance in which each of the six possible outcomes on the pirinola is equally likely to occur. Each player seeks to acquire all the beans or candies.

How to Play

To begin the game, each player must place one bean in a center pile. Turns are taken and play must move to the left. A turn consists of spinning the pirinola and following the directions on the side that is facing up after the spin. The possible resulting directions are listed below.

Pon 1.	Put 1.
Pon 2.	Put 2.
Toma Todo.	Take all.
Toma 1.	Take 1.
Toma 2.	Take 2.
Todos ponen.	Each player puts 1 in the pile.

If *toma todo* turns up, then in order to continue the game, each player must place one bean in a center pile again. Play continues until one player has all the beans or candies.

Note: The rules for toma todo were discussed and followed by Latino students enrolled in the mathematics strategy course for the Emergency Permit Induction Program offered by California State University, Long Beach, 24 February 1998. These teacher-students have played toma todo since childhood.

Rain Bird

Background

The Pueblo Indians were—and are—prolific pottery makers. The Zuni are one of the Pueblo Indian tribes. A design appearing frequently on Zuni water jars is known as the rain bird.

How the term *rain bird* came into existence is unknown; however, Mera (1970) believes that the design's ancestry may be traced to the beginnings of pottery decoration in the Southwest. There have been many variations of the design over the last three centuries.

The Zuni live in western New Mexico. They farm using irrigation. They are noted not only for their pottery but also for basketry, jewelry, weaving, and ceremonial dances.

Materials

paper
pencil or crayon
ruler (optional)
protractor (optional)

Object

The rain bird motif can be used with young students to study the triangle. Older students can construct the different kinds of triangles: equilateral, isosceles, right, and scalene.

Activity

Have the students draw a picture of a triangle. Next the hook-like head, legs, and three or four tail feathers are added. Discuss with students why they think the rain bird design is popular with the Zuni. Discuss the rainfall in the geographical area where the Zuni live.

REFERENCE

Mera, Harry P. *Pueblo Designs*. New York: Dover Publications, 1970.

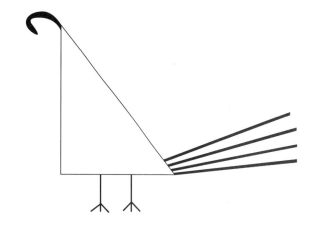

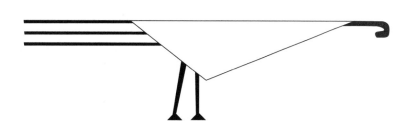

Zia Sun Symbol

Background

This distinctive sun symbol comes from the Zia pueblo of ancient times. To the Zia, four is a sacred number. Four is embodied in the earth's four directions (north, south, east, west); in the year's four seasons (spring, summer, fall, winter); in the day (sunrise, noon, evening, night); and in life itself (childhood, youth, adulthood, old age). According to the Zia, there are four sacred obligations in life. An individual must "develop a strong body, a clear mind, a pure spirit, and a devotion to the welfare of his people." (New Mexico and Santa Fe Visitor Information Center, n.d.)

Today the ancient Zia sun symbol adorns the flag of the state of New Mexico. The Zia sun symbol is red on a background of Spanish yellow.

Materials

Zia Sun Symbol page
compass
ruler with centimeter and millimeter divisions
pencil

Object

Using proportions set by legislative act in New Mexico (New Mexico, 1978). Students will construct the Zia sun symbol. In constructing the symbol students will use a compass, measure in centimeters, and use common fractions (thirds and fifths).

Activity

Distribute the Zia Sun Symbol page and ask students to look carefully at the page and tell what they observe about the symbol. The symbol consists of a circle from which four sets of four straight line segments radiate. These line segments are not all the same length.

Tell students about (or show) the flag of New Mexico. The proportions of the symbol were set by a legislative act. The four sets of line segments, positioned at right angles, have two longer segments in the middle of each set. These longer segments are one-fifth longer than the shorter rays. The circle's diameter is one-third the width of the symbol.

Constructing the Zia sun symbol using the proportions legislated for the flag of New Mexico

1. Have students draw a picture of a circle with a 6-centimeter diameter. (Use a compass setting of 3 centimeters.)
2. Draw the four sets of four line segments. Have the two line segments in the center of each set one-fifth longer than the other two line segments. For example, if the two longer line segments are 6 centimeters, then the two shorter segments are 5 centimeters.

Note: The flag itself has a width that is two-thirds of its length. The Zia sun symbol is one-third of the flag's length.

Additional Activity

The length of the longer line segments in the Zia sun symbol on the New Mexico flag will always equal the diameter of the circle. Ask students to study the symbol and to try to explain why.

A solution (other approaches are possible): If the circle is one-third of the symbol's width, then the length of two sets of longer line segments—those sets opposite each other on the circle—equals two-thirds of the symbol's width.

The length of one set is half that of two sets. Half of two-thirds of the symbol's width is one third of its width, which equals the diameter of the circle.

REFERENCES

New Mexico. *New Mexico Statues, Annotated* (1978), sec. 12-3-2.

New Mexico and Santa Fe Visitor Information Center. "Zia Sun Symbol." n.d.

Zia Sun Symbol

Paiute Walnut Shell Game

Background

This game comes from the Paiute Indians of Pyramid Lake in northwestern Nevada. The Paiutes of this area belong to the group designated as the Northern Paiutes, as opposed to the Southern Paiutes. The Northern Paiutes were scattered over central and eastern California, western Nevada, and eastern Oregon. The Southern Paiutes, who were also known as the Digger Indians because they subsisted on roots, lived in northwestern Arizona, southeastern California, southern Nevada, and southern Utah.

Materials

8 walnut shell halves, filled with pitch and powdered charcoal (In the classroom, clay can be used. Small red and white beads and pieces of abalone shell are inlaid.)
basket tray
small sticks or twigs

Number of players: 2

or more

Object

The Paiute walnut shell game provides practice in one-to-one correspondence and counting to reach an agreed-on sum.

How to play

Place the walnut shell halves in the basket tray and toss. If three or five of the walnut shells land with the flat side up, the player scores one point. No point is scored for any other combination. Each player keeps count of his or her score with small twigs. Players take turns. The winner is the first player to reach an agreed-on sum.

ADDITIONAL READING

Culin, Stewart. *Games of the North American Indians.* 1907. Reprint. New York: Dover Publications, 1975.

Pueblo Fish

Background

The Pueblo Fish design comes from a miniature pottery piece purchased by the author in Pojoaque, New Mexico.

Materials

Pueblo Fish activity page

Object

The Pueblo Fish consists of triangles, some of which overlap. Students will count the number of triangles they see.

Activity

Distribute the Pueblo Fish activity page. Ask students what shapes they see in the body and tail of the fish. Alert them to the possibility that some triangles overlap. Have students determine how many triangles in all. They may be surprised that sixteen triangles can be found in the fish.

Pueblo Fish

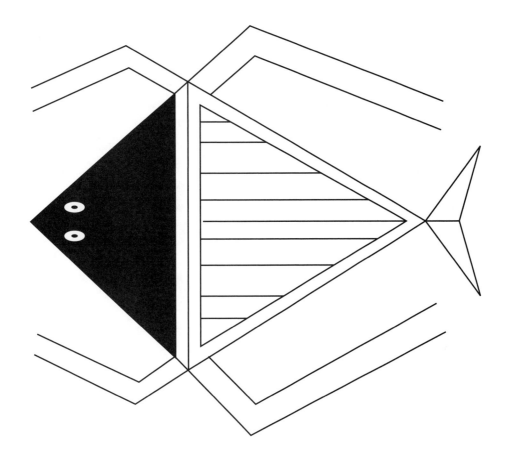

How many triangles do you see?

Tipis

Background

The Cheyenne, Blackfoot, and Sioux American Indian tribes lived on the Great Plains and were hunters. They lived in conical structures called tipis (tepees or teepees). *Tipi* is a word from the Sioux. It is a combination of *ti,* meaning "to dwell or live," and *pi,* meaning "used for." Therefore, a tipi is used to live in.

Tipis are cool during the summer, warm during the winter, and easy to pitch. They withstand high winds, cloudbursts, and they are well lighted and well ventilated. The tipi was made with a frame of poles and was usually covered with buffalo hides. A three-pole foundation was used in the central plains, and a four-pole foundation in the northwest plains. According to Laubin and Laubin (1977), three-pole tipis are stronger than four-pole tipis because the poles are tied together near their tops and provide support for additional poles. Four-pole tipis are not tied this way and have one to four outside guys. Three-pole tipis seldom ever have one guy. During the first decades of the twentieth century, the tipi was no longer in common use on the plains. However, interest in tipis has been renewed and at ceremonies and events during the summer, Native Americans are pitching and occupying them.

Materials

brown construction paper
scissors

Object

Folding paper-chain tipis demonstrates a repeating geometrical pattern. By the use of translation, the student can slide the tipi pattern to the next position.

Activity

Folding paper-chain tipis out of brown construction paper.

1. To make equal divisions, fold paper in half, in half again, and in half again.

2. Open and refold like a fan with accordion folds.

3. Using figure 1 as a guide, draw the tipi on the first folded section, and then slide the pattern on to the next position. Continue to the other end of the paper, being sure to position the middle of the tipi (dotted lines in the figure) on each fold.

REFERENCE

Laubin, Reginald, and Gladys Laubin. *The Indian Tipi: Its History, Construction, and Use.* 2d ed. Norman, Okla.: University of Oklahoma Press, 1977.

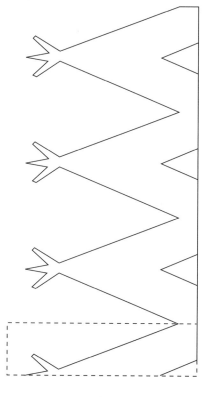

Fig. 1.

Eye-Dazzler

Background

The Navajo (or Navaho) constitute the largest Indian tribe in the United States. During the seventeenth century, these Native Americans lived in the region between the San Juan and Little Colorado rivers in northeastern Arizona. Because their language is related to Athabascan, spoken by their northern neighbors, authorities believe that the Navajo migrated here from farther north.

At one time they raided not only their neighbors, the Pueblo Indians, but also the Spanish and Mexican settlements in New Mexico. However, with the introduction of sheep by the Spanish in the seventeenth century, the Navajo became pastoral, nomadic people.

Navajo women have become well known for their skill in weaving. With the establishment of trading posts in Navajo country, it was not long before blankets became trade items. Since the white settlers did not wear the blankets as the Navajo did, they put the blankets to such other uses as bed and floor coverings. Traders purchased blankets by the pound, and blankets sold in this manner became known as "pound blankets" (Kahlenberg, Hunt, and Berlant 1976).

No two hand-woven Navajo rugs are identical. Some trading post areas have developed rugs with distinctive regional styles. Given graph paper, students may try two other rug styles know as Storm Pattern and Two Grey Hills. Discuss these styles with the students.

Fig. 1.

The development of the storm pattern (fig. 1) has been traced to either 1908 or 1909 (Whitaker 1989). This style has certain essential features. The center rectangle or box is called a "hogan" or "storm house" or the "center of the world." The smaller corner boxes are "houses of the wind." Some interpretations say these represent the four sacred mountains of the Navajo. The four zigzag "lines" connected to the center rectangle represent lightning and have been referred to as "whirling logs." At the top and bottom center of the rug are either stylized or realistic beetle-like figures known as "water bugs" or "pinon beetles." Although the colors used were left to the discretion of the weaver, the colors in the early rugs were black, white, red, and gray. The Tuba City area in Arizona has been known for the weaving of storm pattern rugs.

Two Grey Hills (fig. 2) rugs are well known for their outstanding quality (Crooks and Rogers 1976). The rugs are woven from handspun, natural, undyed wool. The colors in the rugs are black, white, and brown. Also by carding black and white wool together gray is obtained; by carding brown and white wool together, tan is obtained.

Two Grey Hills rugs are named for a village in New Mexico and do not represent hills. These rugs are well balanced and symmetrical. The characteristic geometric designs used include squares, triangles, rectangles, diamonds, steps, crosses, stars, lines, and hooks. Two Grey Hills rugs tend to have a dark border. Rugs may have

Fig. 2.

a "spirit line," a line woven from inside to the edge of the rug to let out the evil spirits. The belief is that the pattern is formed in the weaver's brain. Thus the weaver's thoughts are woven into the rug. The weaver's thoughts must be given a way out so she will not lose her intelligence. Rugs today do not all have a spirit line.

Materials

4-squares-to-an-inch or centimeter graph paper
eye-dazzler design page
ruler

Object

The eye-dazzler design provides practice in reproducing a given pattern. The eye-dazzler blanket style was popular with the Navajo in the 1880s. Note the diamond pattern. Eye-dazzler blankets exploded with vibrant colors. This "blanket" may be colored with red, yellow, and orange.

Students may reproduce a design or create one of their own. See the design on centimeter paper (fig. 3).

Activity

Distribute the eye-dazzler design page and graph paper. Have students reproduce the pattern or create their own "eye-dazzler."

REFERENCES

Crooks, Eleanor, and Rosetta Rogers. "Rugs—Truly Wonderful Navajo Art." In *Our Friends, The Navajos*, pp. 63–85; rev. ed. Of the original book titled *Papers on Navajo Culture and Life* (1970). Tsaile, Ariz.: Navajo College Press, 1976.

Kahlenberg, Mary Hunt, and Anthony Berlant. *The Navajo Blanket.* n.p.: Praeger Publishers in association with the Los Angeles County Museum of Art, 1976.

Whitaker, Kathleen. "The Storm Pattern in Navajo Rugs." *Masterkey* 62 (Winter 1989): 4–9.

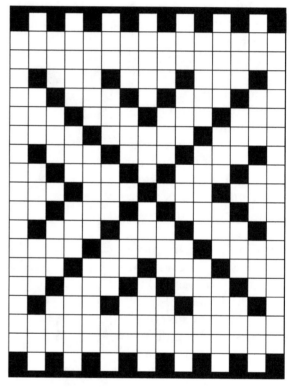

Fig. 3.

Navajo Eye-Dazzler Design

Hogan

Background

The hogan is the traditional dwelling of the Navajo. Mathews (1976) observed that the hogan plays a "significant role in all phases of the religious and secular life of The People." The hogan was mentioned in the creation story, and different types have been constructed throughout its historical development (Kluckhohn, Hill, and Kluckhohn 1971). Various building materials include earth, bark, stone, and logs. The typical hogan's diameter is from eight to 18 feet. It is often hexagonal; however, it may have fewer or more sides. A conical or rounded roof gives the hogan a beehive shape. The entrance of the hogan faces east and the rising sun so the occupants may welcome the new day. Through the years, woven doorway mats were used to prevent the wind or drafts from entering. Later, blanket or lumber doors replaced doorway mats. Inside, the fire or iron stove and smoke hole in the roof are located in the center of the hogan. In cold weather the smoke hole is made larger; in the warmer months the smoke hole is made smaller. A firedrill and hearth made from a block of cottonwood were used for kindling fire. To keep the fire-making equipment dry, it was stored between the roof beams. Many hogans are equipped with wooden poles securely placed horizontally in the roof structure for hanging clothing. Brooms of pinon or juniper branches and fly swatters—wooden frames with woven yucca-leaf wefts—were used. Bundles of grass, cattail, and bark became sleeping mats. The traditional summer home of the Navajo is a brush shelter made from upright poles with a roof of boughs from cottonwood, cedar, or juniper trees.

Materials

box filled with a variety of shapes
glue
Yataahey! activity page

Object

The younger student may construct a hogan picture (fig. 1) by selecting appropriate shapes from the box. The older student may design a floor plan of a hogan or create a floor plan of his own (Yataahey! activity page).

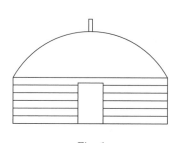

Fig. 1.

Activity

1. Distribute the Yataahey! activity page. Inform students that *Yataahey* is a Navajo greeting similar to the English *hello*. It appears in print spelled various ways.

2. Discuss the directions on the floor plan. Remind students that the door faces east and the fire or iron stove is located in the middle.

3. Have students consider storage areas for cooking utensils; food items such as salt, sugar, and flour; bedding; and clothing.

REFERENCE

Kluckhohn, Clyde, W. W. Hill, and Lucy Wales Kluckhohn. *Navajo Material Culture.* Cambridge, Mass.: Belknap Press of Harvard University Press, 1971.

Mathews, Laura. "The Hogan." In *Our Friends, The Navajos*, pp. 126–30; rev. ed. Of the original book titled *Papers on Navajo Culture and Life* (1970). Tsaile, Ariz.: Navajo College Press, 1976.

Yataahey!

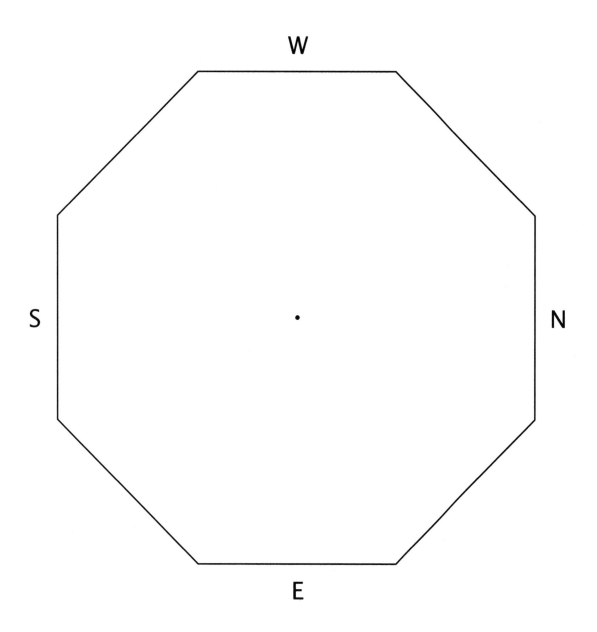

Design a floor plan for your hogan.

Navajo Fry Bread

Background

The Navajo Indians call themselves the *Dînéh*, which means "the people of the earth." Moving with their sheep to find good pasturage, the Navajos did not have electricity in the remote areas where they lived. This fact is reflected in the fry bread recipe—they did not have refrigerators to keep milk fresh and therefore used powdered milk.

Materials

ingredients given below
electric skillet

2 teaspoon baking powder (10 g)
1/2 teaspoon salt (2.5 g)
1/2 cup powdered milk (125 ml)
1 cup warm water

Object

To make Navajo fry bread, students must measure ingredients and follow a set of step-by-step instructions. Note that Navajo cooks use pinches and handfuls. This recipe, given to the writer at Cameron, Arizona, in the Navajo Nation, has been translated into both customary and metric measurements.

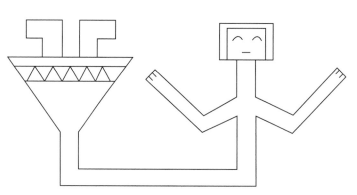

A sandpainting of a rainbow figure, a Navajo deity.

Sometimes a small amount of shortening is added to the ingredients; however, this is usually left out. Mix together the flour, baking powder, salt, and powdered milk. Add warm water to form dough. Knead the dough until it is soft but not sticky. Cover with a cloth and let stand for about 2 hours. Shape the dough into balls about 2" (5 cm) across. Flatten by patting the balls into circles about 8" (20 cm) in diameter. Place the circles in hot shortening about 3 cm deep. Fry at 375°F (190°C) until golden brown on one side, then turn and fry the other side until golden brown.

Activity

Ingredients
2 cups flour (250 g)

Window Rock Design

Background

Window Rock, Arizona, the capital of the Navajo nation, is the seat of tribal government. The elected tribal council meets at least four times during the year. This design, painted turquoise, appears on the facade of the Navajo Arts and Crafts Enterprises building on State Route 264 just east of Navajo Route 12.

Materials

Window Rock Design activity page

Object

This half design using the concept of bilateral symmetry is a mirror reflection about a horizontal line.

Activity

Distribute the activity page and have students complete the design.

Window Rock Design

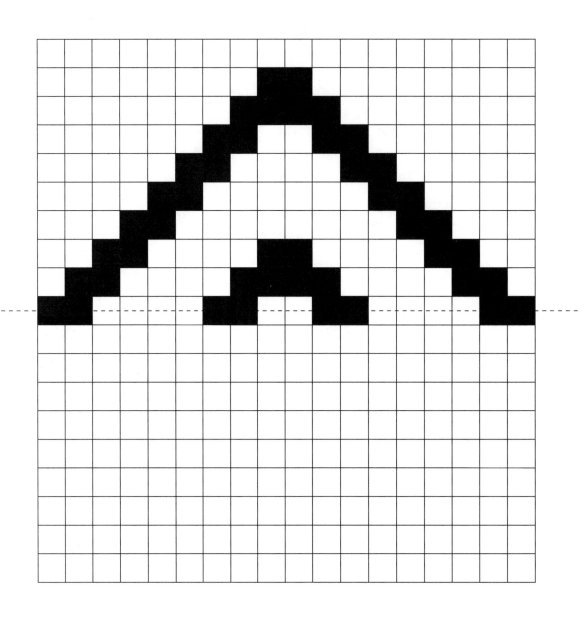

Wasco Deer

Background

In the lower Columbia region of the northwest coast, Wasco women by the nineteenth century made cylindrical baskets decorated with geometric figures of birds, fish, and mammals. They used a twining technique with two colors of weft—one natural and the other dyed dark brown—to create their designs.

Materials

Wasco Deer activity page

Object

The Wasco Deer was created using rectangles. Students will count the number of rectangles that they see. After determining how many rectangles, students may create their own creatures using rectangles.

Activity

Distribute the activity page. Ask students what shape is used in the deer. Have them count the rectangles.

Wasco Deer

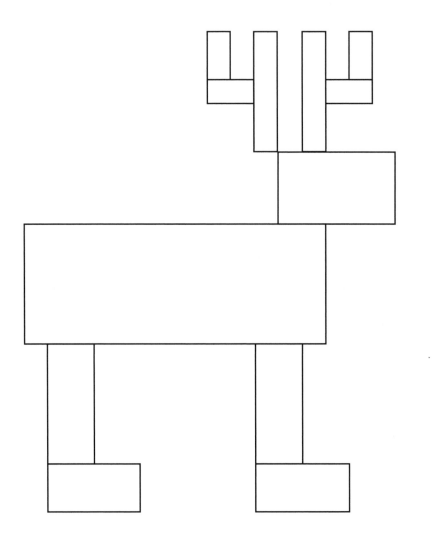

How many rectangles?

Quilts

Background

Quilts—bedcovers made with two layers of cloth—may be filled with cotton, down, or wool. The layers are sewed together in lines or patterns to keep the filling in place. Quilting was a traditional way to bring warmth to fabric in Europe and Asia. Quilts have been used in America since colonial times.

The development of patchwork is usually credited to American pioneer women. These resourceful women created bedcovers from remnants or scraps of worn clothing. At first irregular-shaped fabric pieces were sewn together to form patterns or blocks. These are known as *crazy-quilt patterns*. Crazy quilts were followed by quilts formed by fabric cut in geometric shapes and sewn together to form patterns or blocks. Many of these patterns were named and have been handed down from generation to generation (see fig. 1). Beautiful quilts are made when the block patterns are joined and repeated. The Amish are well known for their quilts.

Materials

Quilt Design activity pages
Quilt Design sheet
colored pencils

Object

Students will repeat traditional geometric block quilt designs. They will work with many geometric shapes—squares, rectangles, diamonds, triangles, hexagons, or octagons. The Quilt Design activity pages provide practice in rotational or turning symmetry.

Activity

Distribute one of the Quilt Design activity pages. Discuss the geometry of the block quilt design. Ask students what happens when the design is turned clockwise. Each design can be turned about its center so that the design coincides with itself. This is *rotational* or *turning sym-*

metry. Have students find the center of rotation of the block quilt design and determine the number of degrees of the rotation. Some of the block quilt designs may have reflection symmetry as well.

Distribute the Quilt Design sheet so that students can create their own block patterns that have rotational symmetry. Ask the students to name their blocks.

Quilters use two basic constructions: (1) a four-patch block, which is a 2×2 square, and (2) a nine-patch block, which is a 3×3 square. The four-patch block is the basic construction used for the Quilt Design activity pages and the Quilt Design sheet. However, the basic 2×2 construction was further divided, giving a 4×4 grid so that diagonals to form triangles can be drawn easily by students.

ADDITIONAL READING

Duke, Dennis, and Deborah Harding, eds. *America's Glorious Quilts*. New York: Park Lane, 1989.

Goldberg, Rhoda Ochser. *The New Quilting and Patchwork Dictionary*. New York: Crown Publishers, 1988.

Quilt names (fig. 1) are listed row by row from left to right:

Row 1
1. Dutchman's Puzzle
2. Ohio Star
3. King's X
4. Wheels

Row 2
5. Windmill
6. Crosses and Losses
7. Anvil
8. Brown Goose

Row 3
9. Flyfoot
10. Puss-in-the-Corner
11. Churn Dash
12. Morning Star

Row 4
13. Flock of Geese
14. Octagon
15. Road to Oklahoma
16. Clay's Choice

Row 5
17. Rail Fence
18. King's Crown
19. Whirlwind
20. Framed Square

Row 6
21. Barbara Frietchie's Star
22. Susannah
23. Cactus Basket
24. Stars and Stripes

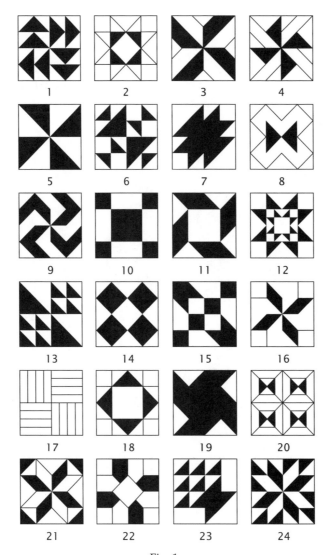

Fig. 1.

Windmill

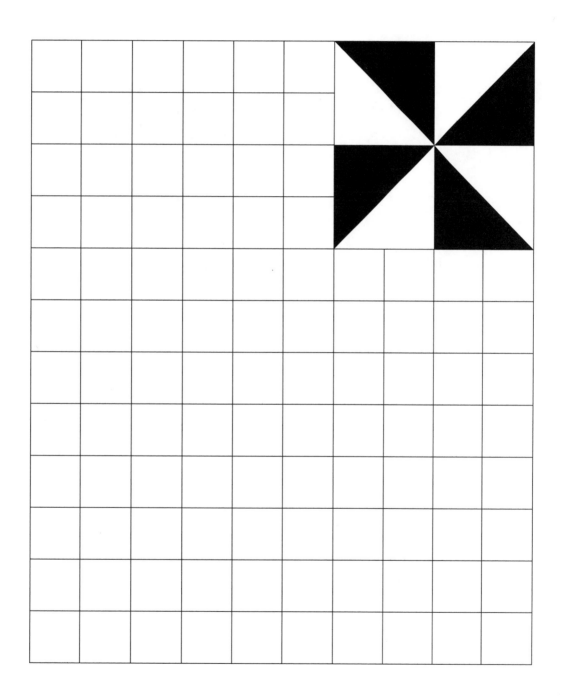

Anvil

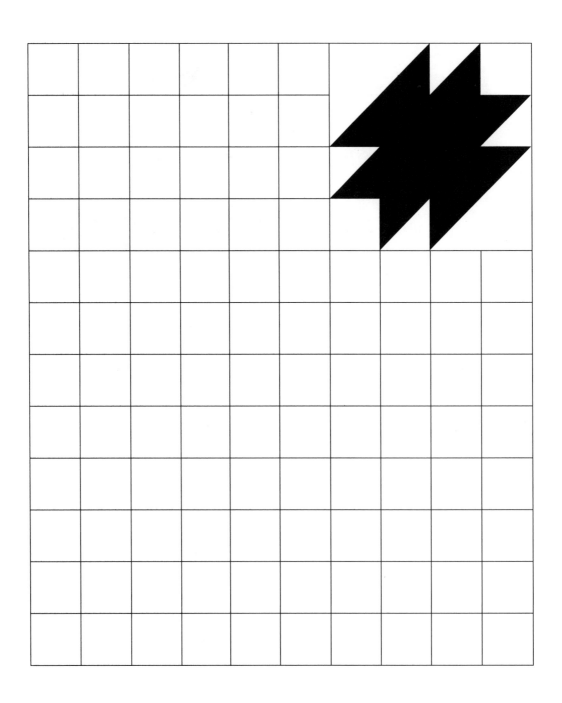

Puss-in-the Corner

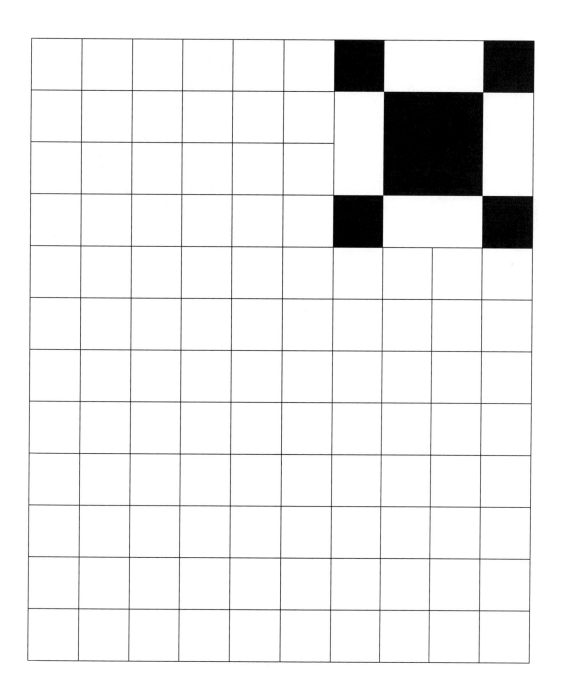

Road to Oklahoma

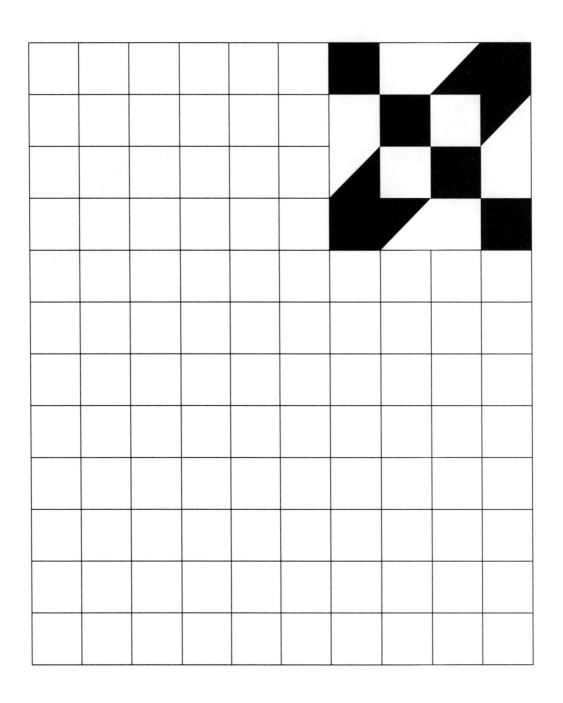

King's Crown

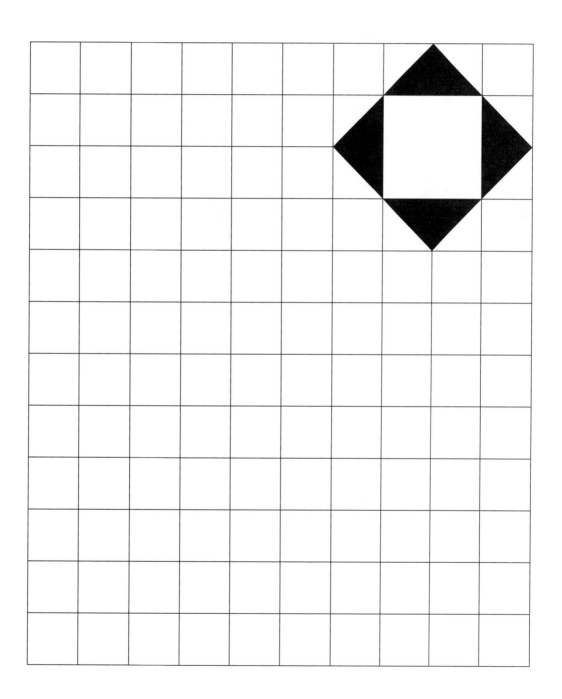

Barbara Frietchie's Star

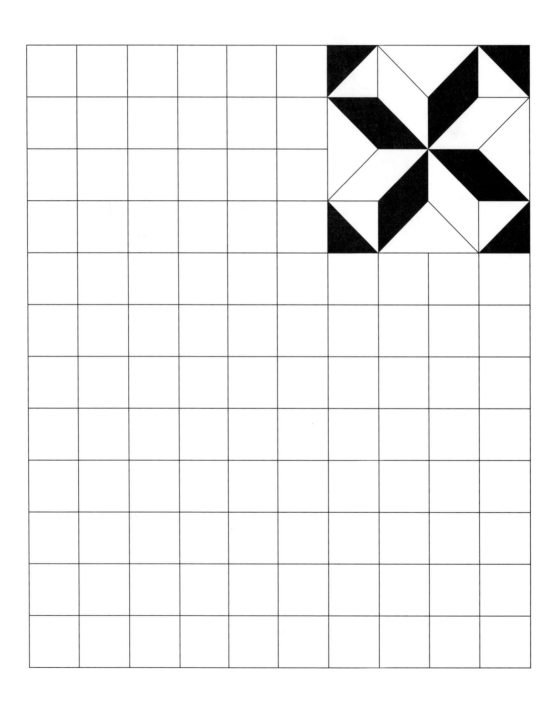

Stars and Stripes

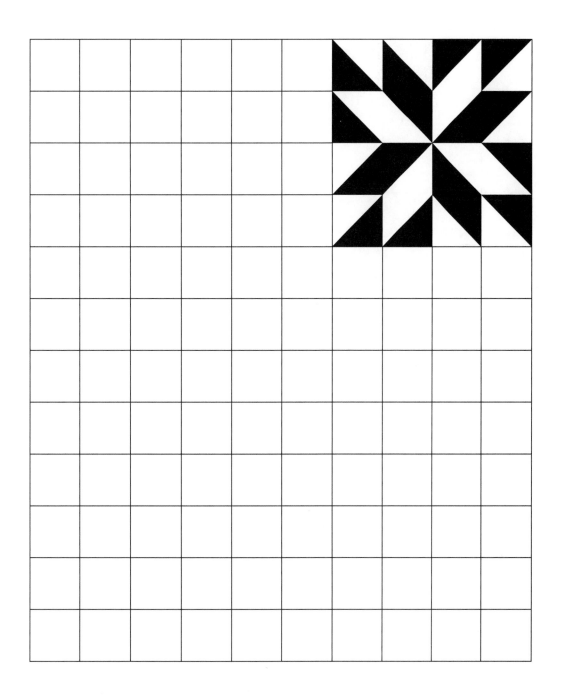

Quilt Design Paper

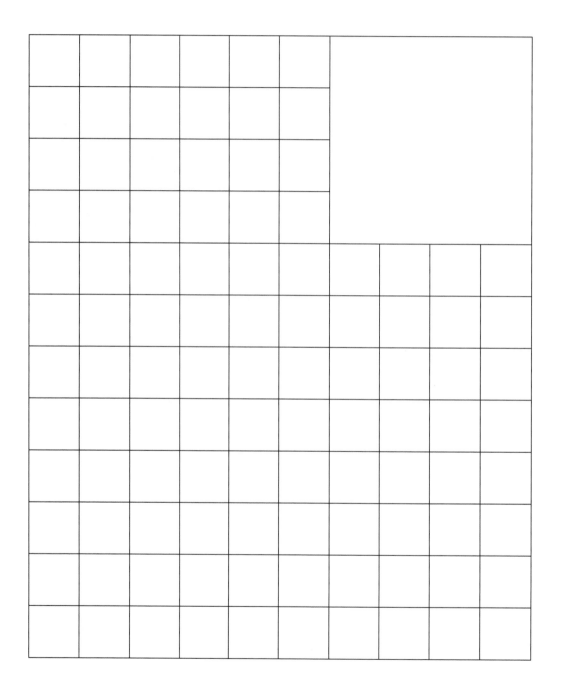

Hex Signs

Background

Pennsylvania Dutch is the name given to people who migrated to eastern Pennsylvania in the eighteenth century. These people were not from the Netherlands, as the name suggests, but rather they were of German descent. In the German language, *Deutsch* means "German." The name was given especially to those who lived in the present counties of Northampton, Berks, Lancaster, Lehigh, Lebanon, York, and in adjacent counties.

Although William Penn established the colony of Pennsylvania as a haven for Quakers, other groups were offered religious freedom. Different sects came, including the Amish and the Mennonites.

The Pennsylvania Dutch have given much to the culture of Pennsylvania. Articles such as their furniture, needlework, pottery, and barns are beautifully decorated. Today their barns are decorated with geometric designs that have attracted the attention of people from afar. Some say that these geometric designs are hex signs to ward off evil spirits, lightning, fire, and sickness. Others say that they are merely barn decorations.

Materials

compass
protractor
ruler
colored pencils

Object

Students are given practice in constructing—and creating—geometric designs. To construct and create their own designs, students need to know basic geometric constructions and their relationship to inscribed polygons.

Fig. 1.

Activity

Have students inscribe a star or rosette of their choice in a circle. Uncircled stars are painted on Pennsylvania Dutch barns; however, they are often circled. The circle is associated with infinity and eternity. Sometimes the way a star is shaded makes it appear to be a more complex design (fig. 1). This is called a "swirling star."

Stars are common hex signs or barn decorations. They may have four, five, six, seven, eight, ten, twelve, sixteen, or thirty-two points. Various meanings have been associated with the number of star points: four, good luck (with raindrops added, the symbol for rain); five, good luck and the five senses; six, love; eight, goodwill and abundance; twelve, knowledge and wisdom; and sixteen, rationalism and justice.

Constructing a six-petal rosette

1. Draw a circle having a radius of 5–8 centimeters with a compass.

2. Do not change the compass setting.

3. Place the point of the compass on the circumference of the circle.

4. Make a small arc on the circumference (fig. 2).

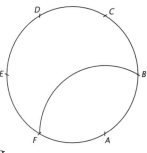

Fig. 2.

117

5. Continue around the circumference making small arcs. Label these arcs *A* through *F*.

6. Place the point of the compass on *A* (where the small arc intersects the circumference) and draw arc *FB*.

7. Uses each small arc intersection as a point of reference. Continue around the circumference

from *B*	arc *AC*
from *C*	arc *BD*
from *D*	arc *CE*
from *E*	arc *DF*
from *F*	arc *EA*

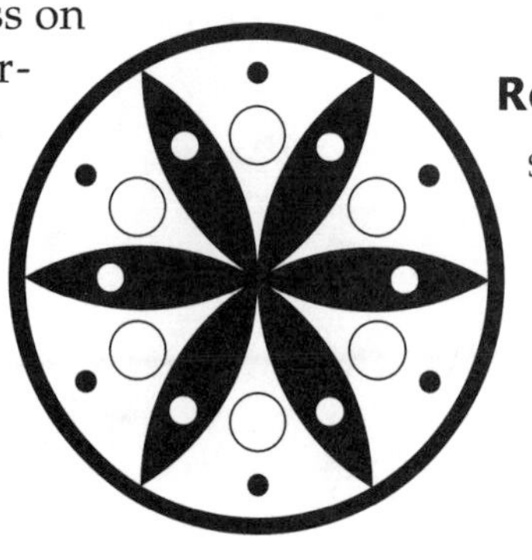

8. The rosette is complete and ready to be embellished (figs. 3, 4).

Reference

Smith, Elmer L. *Hex Signs and Other Barn Decorations.* Lebanon, Pa.: Applied Arts Publishers, 1978.

Fig. 3.

Fig. 4.

118

Hopi Rain Cloud

Background

In northeastern Arizona, three rugged promontories project from the south side of Black Mesa, a great circular upland with a diameter of about ninety-six kilometers. To the Hopi, Black Mesa is *Tuuwanasavi,* "Center of the Universe." It is where the migrations of the Hopi clans met and settled about a thousand years ago. These migrations had taken the Hopi in the four directions to the ends of the land. Following the instructions of Maasaw, guardian spirit of the Hopi Fourth World, some clans turned north while others turned south. Retracing their routes, they then turned east and west and returned again.

The rain cloud symbol is frequently used by the Hopi (Fewkes 1892). It may be placed over altars and on roof slabs, on kachinas (dolls), pottery wares, and even gaming implements. This is not surprising, because Hopi existence at one time depended on rainfall. Farming is as important to the Hopi today as it was to their ancestors. Farming in an arid region, they grow beans, corn, melons, squash, and cotton. Corn represents the mother of life.

Materials

activity page or other paper
compass
ruler

Object

The Hopi Rain Cloud activity involves the construction of semicircles, or half circles. The prefix *semi-* signifies "half"; a semicircle is half a circle.

Activity

Have students first look at a circle with diameter $\overline{AB}$. If the points A and B are the endpoints of the diameter, each of the two arcs—above and below—is called a "semicircle". The arc degree measurement of a semicircle is 180. This is written as $m(\widehat{AB}) = 180$.

Constructing the rain-cloud symbol

1. Use the activity page or draw a line segment in about the middle of the paper. Select a length that is divisible by four.

2. Mark and label points A, B, and C at the following locations (see fig. 1):

   A—one-fourth of line segment from left endpoint

   B—one-fourth of line segment from right endpoint

   C—midpoint of the line segment and above it at a distance that is one-fourth of the line segment

3. Place the point of the compass at A and twist the compass to construct a semicircle above the line. Repeat at B and C (see fig. 2).

4. Draw short vertical line segments for rain (see fig. 2).

5. Draw "lightning snakes" from the clouds if desired (see fig. 3).

REFERENCE

Fewkes, J. Walter. "A Few Summer Ceremonials at the Tusayan Pueblos." In *A Journal of American Ethnology and Archaeology,* vol. 2, pp. 1–160, edited by J. Walter Fewkes. New York: Houghton, Mifflin & Co., 1892.

Hopi Rain Cloud

Fig. 1.

Directions

1. Place the point of your compass at *A* and twist the compass to construct a semicircle above the line. Repeat at *B* and *C* (see fig. 2) These semicircles are the rain clouds.

2. Draw short vertical line segments for rain (see fig. 2)

3. Draw "lightning snakes" from the clouds if desired (see fig. 3).

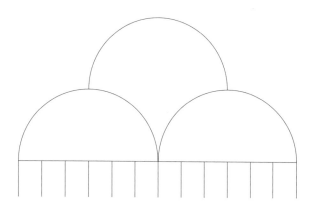

Fig. 2.

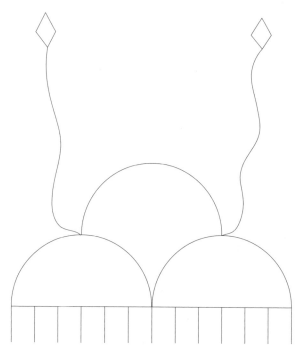

Fig. 3.

Pottery Designs

Background

Hopi pottery making began about fifteen hundred years ago in the Four Corners area where Arizona, Utah, Colorado, and New Mexico meet. From the beginning, the Hopi decorated their pottery with geometric designs. The following designs were found on pottery ware recovered from the ruins of Sikyatki, an ancient pueblo. The destruction of Sikyatki took place before 1540. The cause of the destruction is unknown. Sikyatki was excavated in 1895 by the Smithsonian Institution under the direction of J. Walter Fewkes. Fewkes ([1898] 1973) referred to this activity as the "Expedition to Arizona in 1895."

The elaborately decorated pottery that combined life forms and geometric designs is known today as Sikyatki Polychrome. Fewkes considered this pottery the "most highly developed" of Hopi ceramics.

Materials

Pottery Designs activity page
ruler

Object

The Pottery Designs activity page provides practice in completing repeating patterns. These patterns were bands encircling the exterior of Sikyatki food bowls. Mathematically, the patterns are obtained by a combination of translation—simply sliding to the next position—and reflection on an axis.

Activity

1. Pottery Design, activity 1

As each band is studied, discuss with students the geometrical shapes and line segments of the design. Ask if translation, reflection, or a combination of both will be used to obtain the completed pattern.

Cardboard patterns can be cut out using the two-band shapes. The top band has alternating right triangles that meet at their right angles, and the lower band has two isosceles triangles meeting at their top vertexes. Students then will be able to slide and flip the patterns to check their answers.

2. Pottery Designs, activity 2

The composition of this exterior band is more complex than the bands in activity 1. The overall form is rectangular; however, three right triangles, another rectangle, and five sets of vertical parallel bands are within the rectangular band.

To complete the top band, have each student flip the pattern mentally first. The completed design is a Sikyatki food-bowl decoration.

For the lower band, pose the following question: "What would the band decoration look like if the Hopi pottery maker had decided not to flip the pattern?"

REFERENCE

Fewkes, Jesse Walter; *Designs on Prehistoric Hopi Pottery.* 1898. Reprint. New York: Dover Publications, 1973.

Solution

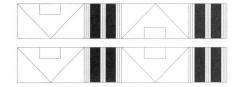

Pottery Designs

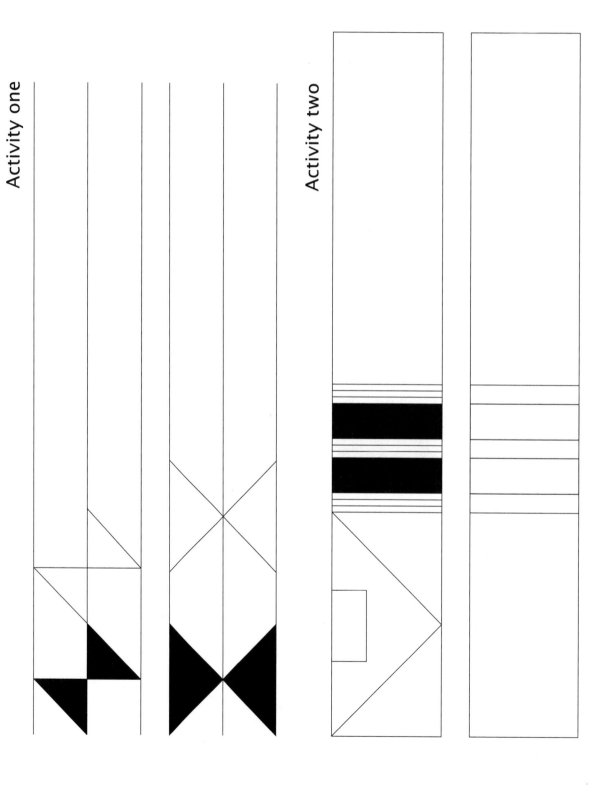

Activity one

Activity two

Hopi Bird

Background

The Hopi Indians of northeastern Arizona are well known for their beautiful pottery ware, which is decorated lavishly with geometric shapes and designs. The bird figure that follows was found on a piece of prehistoric ware from ruins on the Little Colorado River (Fewkes [1898] 1973).

Hopi pottery making continues to flourish today. Hopi potters use the coil method to make their pottery. Ropes of clay are spiraled, one on top of another. The pot is smoothed, dried, and coated with watery clay. Then the pot is polished with a smooth stone. The design is carved or painted on the pot before firing.

Materials

Hopi Bird activity page

Object

The Hopi Bird activity page can be used with students to study geometric shapes—circles, triangles, rectangles, squares, and trapezoids. Parallel lines can also be noted.

Activity

Distribute the activity page. Explain to students that this is a figure found on a piece of prehistoric pottery made by the Hopi. Ask what living thing they think this figure represented and why. Ask students to point out the different geometric shapes they see in the Hopi Bird.

REFERENCE

Fewkes, Jesse Walter. *Designs on Prehistoric Hopi Pottery.* 1898. Reprint. New York: Dover Publications, 1973.

Hopi Bird

Sky Window

Background

The Hopi sky god, Co-tuk-i-nung-wa, is represented by a five-pointed star. The star symbolizes the deity's eye, through which he watches all things. The eye of Co-tuk-i-nung-wa is also known as the *sky window*. Through the sky window, the sky god gives all things vitality.

Materials

paper
compass
protractor

Object

The sky window can be used as a compass-and-protractor activity. Students inscribe a pentagon in a circle and construct a pentagram, or star.

The sky window here is drawn from a Polacca Polychrome jar, 1820–1860, on display in the Heard Museum's exhibition, "America's Great Lost Expedition: The Thomas Keam Collection of Hopi Pottery from the Second Hemenway Expedition, 1890–1894." The Heard Museum is in Phoenix, Arizona.

Activity

Inscribing a pentagon inside a circle (fig. 1)

1. Draw a circle having a radius of 5–8 centimeters with a compass.
2. Label the radius $\overline{AB}$.

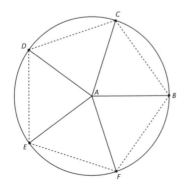

Fig. 1.

3. Divide the circle (360°) into five equal parts of 72° each and mark these points.
4. Draw connecting lines from the center to the points, constructing "angles" of 72° (solid lines in figure).
5. Label each new radius drawn, $\overline{AC}$, $\overline{AD}$, $\overline{AE}$, and $\overline{AF}$.
6. Draw line segments (dotted lines in figure) from B to C, C to D, D to E, E to F, and F to B to form a pentagon.

Constructing a pentagram or star (fig. 2).

Knowledge of the pentagon will enable students to construct the pentagram.

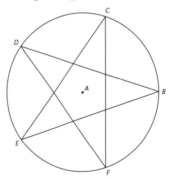

Fig. 2.

1. Divide the circle into five equal parts and mark these dividing points as before.
2. Draw line segments from B to D, B to E, C to E, C to F, and D to F.
3. Erase the circle and the line segments inside the pentagon.

ADDITIONAL READING

The Heard Museum. *America's Great Lost Expedition: The Thomas Keam Collection of Hopi Pottery from the Second Hemenway Expedition, 1890–1894.* Phoenix, Ariz.: Heard Museum, 1980.

Sun Kachina Mask

Background

Hopi kachinas are believed to be supernatural beings embodying the spirits of living things or of dead ancestors. The Hopi men who don masks and impersonate the supernatural beings in the kivas (underground structures) and plazas of the villages are also called *kachinas*. The Hopis believe that when a man impersonates a kachina he acquires the kachina's spirit.

The Sun Kachina, also called *Tawa Kachina*, represents the spirit of the Sun God. The mask is a yucca basket with a forehead that is red on the left side and yellow on the right side. The remaining face is painted green. Surrounding the face are eagle feathers.

Materials

Sun Kachina Mask activity page
mirrors (optional)

Object

The Sun Kachina Mask activity provides practice in mirror geometry—that is, this is an activity in symmetry with respect to a straight line. The right half of the mask is to be a mirror image of the other half, an example of bilateral symmetry. *Bi-* means "two," and *lateral* refers to the sides. To say it simply, the two sides will match.

Activity

Distribute the activity page. Discuss the meaning of kachinas and ask students why they think the sun commands such importance. Consider where the Hopi live—in a semiarid region of northeastern Arizona.

Have students place a mirror perpendicular to the line in the middle of the mask. This line divides the mask into two congruent parts. The missing part on the right side can be obtained from the part on the left. After the students have studied the effect of the mirror, ask them to complete the mask. Clearly, this is an activity for older students and not a beginning activity for young students. Even so, time spent experimenting with mirrors to obtain mirror images would be well spent.

Sun Kachina Mask

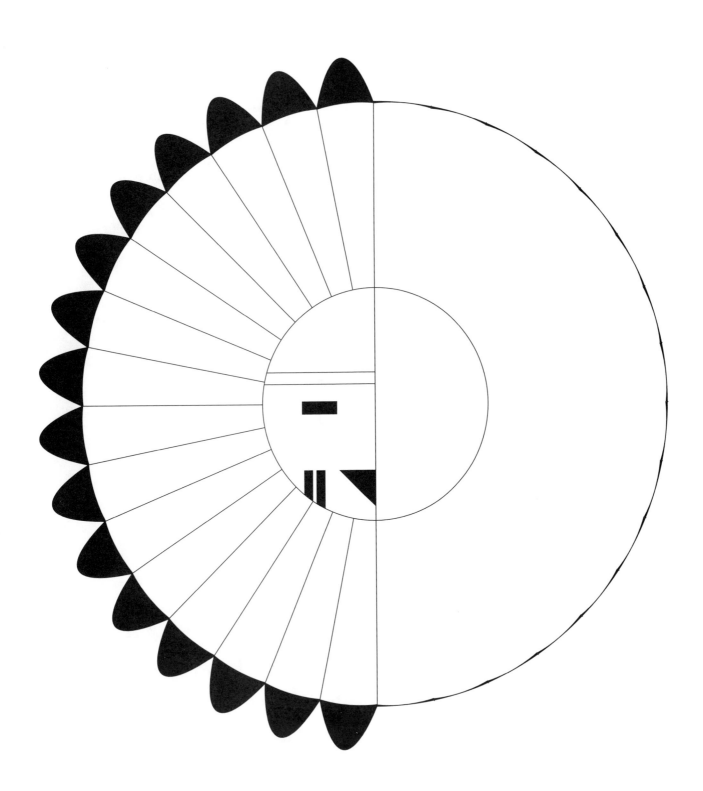

Additional Activity Pages

Background

Sikyatki Butterfly Design

For the Hopi, the butterfly is a symbol of fruitfulness. Butterfly or moth figures appear on the pottery unearthed in 1895 at the ruins of the ancient pueblo Sikyatki. Anthropologist J. Walter Fewkes ([1898] 1973) identified five typical butterfly or moth figures found on Sikyatki pottery. An element common to all was a triangular body.

Without question, the most beautiful figure of all is the butterfly design on the famous "butterfly vase." The figure is repeated six times, which probably represents a relationship to the cardinal points—north, west, south, east, above (or zenith), and below (or nadir).

The six butterfly heads face the opening of the vase. All the butterflies do not have the same head markings but alternate in the following manner: A butterfly having a cross on the circular head with a dot in each section is alternated with a butterfly having only dots on the head and no cross. The former represents a female butterfly and the latter represents a male butterfly. North corresponds to male, west to female, south to male, east to female, above to male, and below to female.

Hopi Turtle

Hopi basketry items are found less frequently in stores and galleries than other Hopi artistic forms, such as kachina dolls, pottery, and silver-overlay jewelry. Basketry items, especially plaques, not only fulfill utilitarian purposes but also serve as gifts or paybacks on special social occasions. Baskets are woven in addition to the many everyday responsibilities in a Hopi woman's life. A major daily responsibility relates traditionally to procuring and preparing food (Teiwes 1996).

Hopi weavers use three different weaving techniques: plaiting, coiling, and wicker weaving. The wild plants providing the natural fibers for basket weaving are gathered nearby on the mesas. The automobile enables weavers to travel greater distances to gather suitable leaves. Coiled baskets and plaques are made from the leaves of yucca plants gathered at certain times of the year to secure particular natural colors. Five colors are available to the weaver: white, green, yellow, black, and red. Yucca leaves must be dyed to obtain black and red.

The Hopi Turtle activity page comes from a coiled plaque in the author's collection. The coiled plaque was purchased at Shungopovi on Second Mesa.

Materials

Sikyatki Butterfly Design activity page
Hopi Turtle activity page
mirrors (optional)

Object

These half designs provide additional practice using the concept of bilateral symmetry, or symmetry about a line.

Activity

Distribute an activity page and let students make their own unhurried investigations with mirrors. Ask if the design is bilaterally symmetric to the line. Have students complete the design.

REFERENCES

Fewkes, Jesse Walter. *Designs on Prehistoric Hopi Pottery*. 1898. Reprint. New York: Dover Publications, 1973.

Teiwes, Helga. *Hopi Basket Weaving: Artistry in Natural Fibers*. Tucson, Ariz.: The University of Arizona Press, 1996.

128

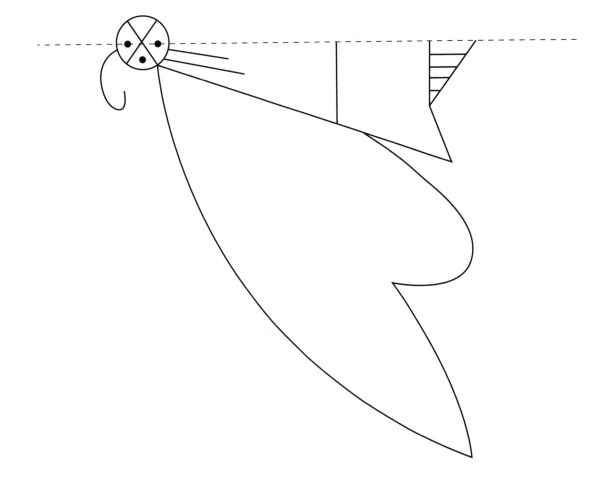

Sikyatki Butterfly Design

Hopi Turtle

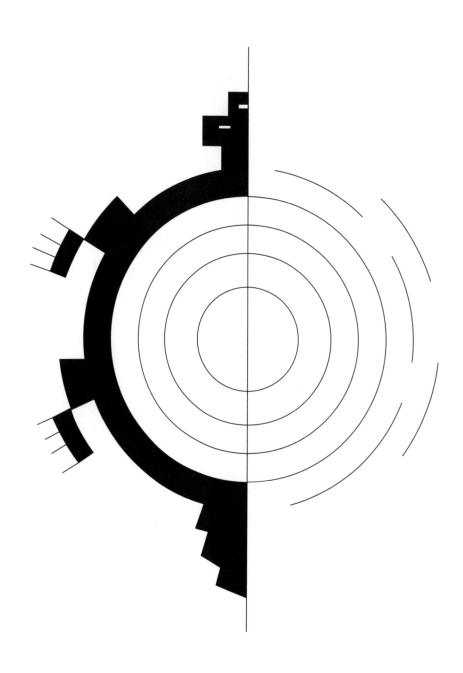

Hemis Kachina

Background

Hopi men carve kachina dolls from dry roots of dead cottonwood trees, which they find along the banks of the Little Colorado River or along one of the washes near the three Hopi mesas. After cutting the cottonwood to the length desired and whittling the piece of wood into the desired shape of the kachina, the Hopi craftsman smooths the doll with a wood rasp and sands it with a piece of sandstone. A nose, horns, ears, and the headdress, or *tableta,* are added with small wooden pegs or glue. The kachina doll is then ready for painting and final adornment with feathers.

Kachinas—Hopi men masked to impersonate supernatural beings—present kachina dolls to Hopi children at kachina ceremonies. The dolls are not dolls in the usual sense. They serve an educational purpose: they help the children learn what the many different kachinas look like. The children also learn about the rituals.

Materials

graph paper
instructions

Object

The Hemis Kachina activity page provides practice in plotting given points and drawing line segments from point to point to reveal a hidden picture.

Activity

The kachina of the picture graph represents the Hemis Kachina, also called the "Jemez Kachina." It is frequently (and mistakenly) called the "Niman Kachina" because it is seen most often in the Niman or Home Dance.

The picture graph was drawn from a Hemis kachina doll purchased in 1928 by a member of the author's family at Keams Canyon, Arizona. The picture is drawn in proportion—head to body, tableta (headdress) to body, and legs to body. Further, the picture graph depicts the old style of the kachina doll: The arms were carved against the body. Today a kachina doll may be carved so that the kachina doll appears to be dancing.

The carved kachina doll is elaborately decorated, especially the tableta; however, for our graphing purposes the doll has enough detail to be recognizable in a group of kachina dolls as a Hemis Kachina. After completing the picture graph, students may wish to draw feathers at the top and corners of the tableta.

ADDITIONAL READING

Colton, Harold S. *Hopi Kachina Dolls with a Key to Their Identification.* Albuquerque, N.M.: University of New Mexico Press, 1959.

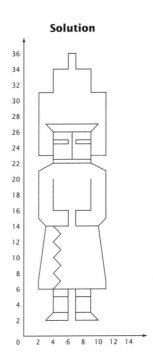

Solution

131

Hemis Kachina

Plot these points on your graph paper:

(6,36)	(7,36)	(4,34)	(6,34)	(7,34)	(9,34)
(2,31)	(4,31)	(9,31)	(11,31)	(3,27)	(10,27)
(4,26)	(6.5,26)	(9,26)	(4,25)	(6,25)	(7,25)
(9,25)	(4,24.5)	(6,24.5)	(7,24.5)	(9,24.5)	(2,23)
(4,23)	(9,23)	(11,23)	(4,22.5)	(6.5,22.5)	(9,22.5)
(4,22)	(9,22)	(2,21)	(11,21)	(4,20)	(9,20)
(4,16)	(6,16)	(7,16)	(9,16)	(2,15)	(11,15)
(3,14)	(4,14)	(6,14)	(7,14)	(10,14)	(5,13)
(4,12)	(5,11)	(4,10)	(5,9)	(4,8)	(2,7)
(5,7)	(11,7)	(2,6)	(4,6)	(6,6)	(7,6)
(9,6)	(11,6)	(4,5)	(6,5)	(7,5)	(9,5)
(4,3),	(6,3)	(7,3)	(9,3)	(3,2)	(6,12)
(7,2)	(10,2)				

Draw line segments from—

1. (4,23) to (2,23)	24. (2,15) to (3,14)	47. (10,2) to (9,3)
2. (2,23) to (2,31)	25. (3,14) to (6,14)	48. (9,3) to (9,6)
3. (2,31) to (4,31)	26. (6,14) to (6,16)	49. (4,5) to (6,5)
4. (4,31) to (4,34)	27. (6,16) to (4,16)	50. (4,3) to (6,3)
5. (4,34) to (6,34)	28. (4,16) to (4,20)	51. (7,5) to (9,5)
6. (6,34) to (6,36)	29. (9,22) to (11,21)	52. (7,3) to (9,3)
7. (6,36) to (7,36)	30. (11,21) to (11,15)	53. (4,14) to (5,13)
8. (7,36) to (7,34)	31. (11,15) to (10,14)	54. (5,13) to (4,12)
9. (7,34) to (9,34)	32. (10,14) to (7,14)	55. (4,12) to (5,11)
10. (9,34) to (9,31)	33. (7,14) to (7,16)	56. (5,11) to (4,10)
11. (9,31) to (11,31)	34. (7,16) to (9,16)	57. (4,10) to (5,9)
12. (11,31) to (11,23)	35. (9,16) to (9,20)	58. (5,9) to (4,8)
13. (11,23) to (9,23)	36. (3,14) to (2,7)	59. (4,8) to (5,7)
14. (9,23) to (9,22)	37. (2,7) to (2,6)	60. (5,7) to (4,6)
15. (9,22) to (4,22)	38. (2,6) to (11,6)	61. (6.5,26) to (6.5,22.5)
16. (4,22) to (4,26)	39. (11,6) to (11,7)	62. (4,22.5) to (9,22.5)
17. (4,26) to (3,27)	40. (11,7) to (10,14)	63. (4,25) to (6,25)
18. (3,27) to (10,27)	41. (4,6) to (4,3)	64. (6,25) to (6,24.5)
19. (10,27) to (9,26)	42. (4,3) to (3,2)	65. (6,24.5) to (4,24.5)
20. (9,26) to (9,23)	43. (3,2) to (6,2)	66. (9,25) to (7,25)
21. (4,26) to (9,26)	44. (6,2) to (6,6)	67. (7,25) to (7,24.5)
22. (4,22) to (2,21)	45. (7,6) to (7,2)	68. (7,24.5) to (9,24.5)
23. (2,21) to (2,15)	46. (7,2) to (10,2)	

Totolospi

Background

Totolospi, pronounced to-tó-los-pi, is a Hopi game of chance that was played with two or three cane dice and a counting board inscribed on stone. Totolospi was played by adults as well as children.

Interestingly, totolospi has been described as a game of fox and geese and as resembling checkers, Monopoly, and Parcheesi (Watkins 1944). These diverse descriptions occur as the result of the variety in the stone totolospi counting boards acquired by different collectors.

Three varieties of totolospi counting boards are included: (1) an ellipse for two players, (2) two ellipses intersecting to form a cross for four players, and (3) a rectangular board for two or four players. The rectangular board is about the size of the inscribed red sandstone slab at the Southwest Museum in Highland Park, California. This board was drawn after measuring the board through the glass of the museum case. The cases are permanently sealed to protect the artifacts.

Materials

counting board

2 or 3 cane dice (Half-round dowels can be used for the cane dice. Cut each 9.5 cm in length. If desired, paint the round sides white and the flat sides red. If half-round dowels are not available, use Popsicle sticks.)

2 or 4 counters called "animals" (The Hopi used kernels of colored corn for the animals. Each player has a different color.)

Number of players: 2 or 4

Object

Totolospi is a game of chance. With older students, consider possible outcomes when two or three cane dice are used. Students can use tree diagrams as a useful tool for calculating probabilities. For younger students, the game provides practice in counting by ones and twos on the elliptical boards and by fives and tens on the rectangular board.

How to play

1. *Totolospi for two:* Elliptical counting board

Each player places a counter (or animal) on the nearest circle. Alternate turns are taken. Movement is on the line segments of the board, not on the spaces between the segments. The moves are determined by holding three cane dice in one hand and dropping them on end. If three round (white sides) land up, the player advances two lines. If three flat (red sides) land up, the player advances one line. If a combination is thrown, the player loses his or her turn. The player reaching the opposite side first wins.

2. *Totolospi for four:* Cross-shaped counting board

Each player places an animal on the circle of the arm of the cross nearest to him or her. Players take turns. If three round sides land up, the player advances two lines. If three flat sides land up, the player advances one line. If a combination is thrown, the player loses his or her turn. The winner is the first to reach the opposite end of the arm he or she is playing on.

3. *Totolospi for two or four:* Rectangular counting board

133

For this version, movement is on the spaces rather than the lines. Players place their animals in the center circle, or "the house," the circle at the end of the strip within the rectangle. Movement is determined by dropping two cane dice. If both round sides land up, the player moves ten spaces. If both flat sides land up, the player moves five spaces. A player continues to play as long as the cane dice show both sides alike. If the dice land with unlike sides, then the next player plays. Two animals may not occupy the same space. The second animal landing on a space sends the, first one back to the house to begin again.

At the end of the inside path, the player must turn and continue around the board, called "the home road." The player may turn either right or left. At the last corner after going around the board, if the player's dice land flat sides up, he or she may go to the door of the house and later proceed inside to the starting point. If, however, the player's dice land round sides up, he or she has gone past the door and must go around the home road again. The winner is the first player to return to the starting place. If, after a second trip around the board, the player goes past the door of the house again, he or she loses.

After playing

With older students, consider the possible outcomes when two or three cane dice are used. To begin, if one cane is dropped, it may land with the round side up (R) or the flat side up (F). The possible outcomes of dropping two cane dice are shown in figure 1.

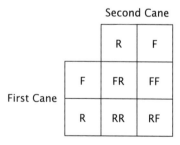

Second Cane

		R	F
First Cane	F	FR	FF
	R	RR	RF

Fig. 1.

There are 2×2 possibilities. Since each outcome is equally likely to occur, the probability of a given event P can be calculated as follows:

$$P \text{ (an event)} = \frac{\text{number of outcomes players are interested in}}{\text{total number of possible outcomes}}$$

Therefore, the probability of the cane dice landing with two like sides up is 2 out of 4. This is written 2/4, or 1/2. The probability of 1/2, or 0.5, means that the chance of the dice landing with two sides alike is equal to the chance of the dice not landing with two sides alike.

However, in the rectangular totolospi board, the player needs two round sides up to move ten spaces. The probability of the cane dice landing with two round sides up is 1 out of 4, or 1/4. If the player should need five to go to the door of the house, the probability is the same. The probability of two flat sides landing up is 1 out of 4, or 1/4.

For the other two board games—the ellipse and the cross—three cane dice are used. The possible outcomes are listed in the tree diagram (see fig. 2). There are $2 \times 2 \times 2$ possibilities.

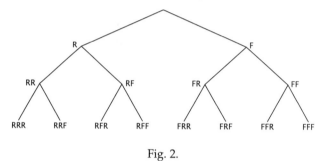

Fig. 2.

Movement in these totolospi games is dependent on the dropping of three cane dice with three round sides landing up or three flat sides landing up. These events are mutually exclusive; that is, if all three land round sides up, this excludes three flat sides up or other outcomes.

The probability of three round sides landing up is 1/8. The probability of three flat sides landing up is 1/8. The probability of either three round or three flat sides landing up is determined as follows:

$$P \text{ (R or F)} = P \text{ (R)} + P \text{ (F)}$$
$$= 1/8 \quad + 1/8$$
$$= 2/8$$

The probability of either three round or three flat sides landing up is 2/8, or 1/4. Ask students if they think that three cane dice landing with three round sides up or with three flat sides up is easy or difficult to obtain.

REFERENCES

Watkins, Frances E. "Indians at Play: I—Hopi Parcheesi." *Masterkey* 18 (September 1944): 139–41.

Totolospi

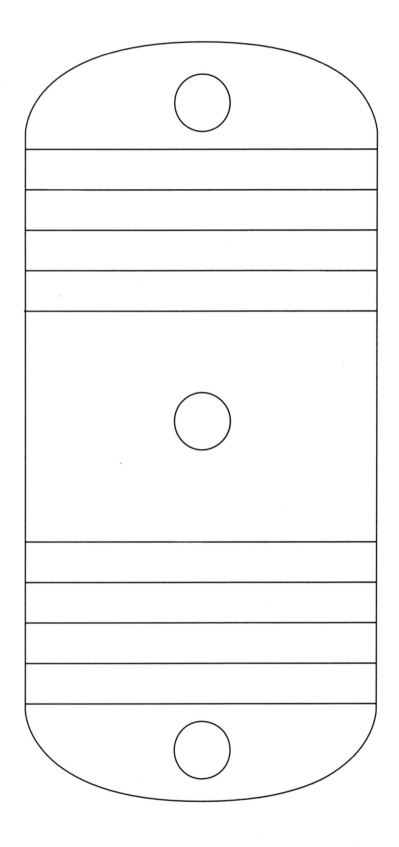

Totolospi

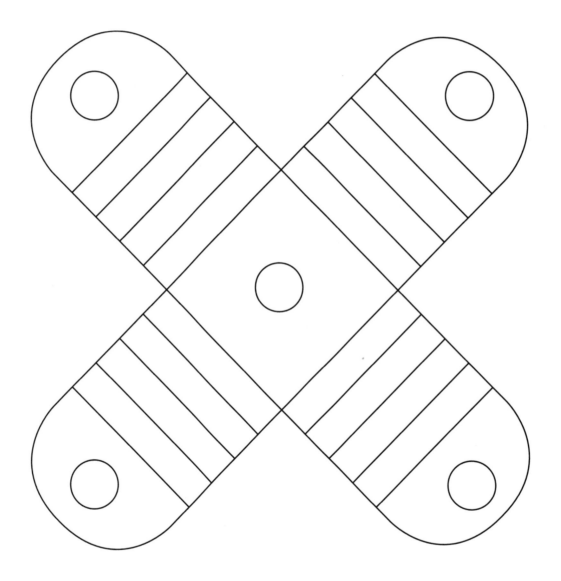

Totolospi

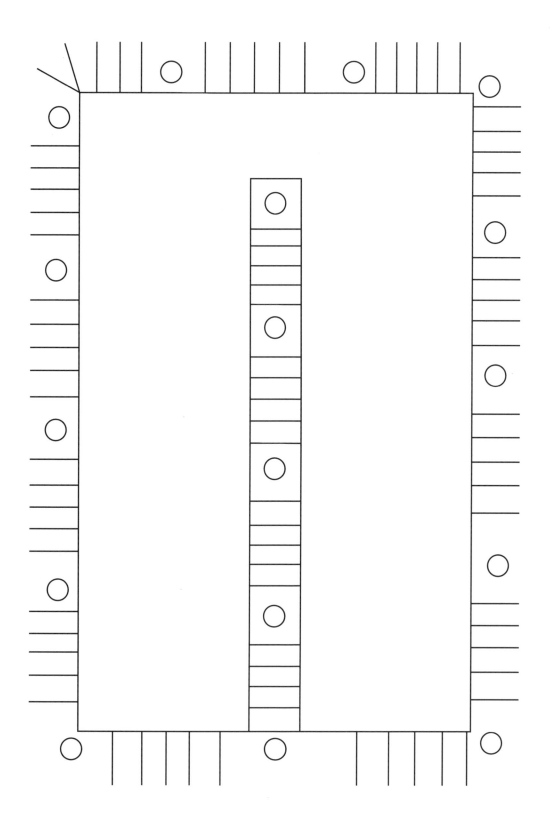

AFRICA

GRADES

ACTIVITY/GAME	1	2	3	4	5	6	7	8
Egyptian Match				●	●	●		
Egyptian Numeration system				●	●	●		
Pyramid					●	●	●	●
Famous Dates				●	●	●		
Senet						●	●	●
Twenty Squares					●	●	●	●
Wari			●	●	●	●	●	●

ASIA

GRADES

ACTIVITY/GAME	1	2	3	4	5	6	7	8
Ko-no			●	●	●	●		
Five-Field Ko-no			●	●	●	●		
Nyout				●	●	●	●	●
Tangram			●	●	●	●		
Origami Cup			●	●	●	●	●	●
Magic Squares					●	●	●	●

OCEANIA

GRADES

ACTIVITY/GAME	1	2	3	4	5	6	7	8
Lu-lu	●	●	●	●	●	●		
Konane				●	●	●	●	●
Petroglygh	●	●	●	●	●	●	●	●
Tapatan	●	●	●	●	●	●	●	●
Tablita			●	●	●	●	●	●
Taniko Patterns				●	●	●	●	●
Mu Torere				●	●	●	●	●

EUROPE

GRADES

ACTIVITY/GAME	1	2	3	4	5	6	7	8
Nine Men's Morris	●	●	●	●	●	●	●	●
Julekurv				●	●	●	●	●
Celtic Design				●	●	●	●	●
Erin Go Bragh					●	●		
Roman Numerals				●	●	●	●	●
Asalto				●	●	●	●	●

THE MIDDLE EAST

GRADES

ACTIVITY/GAME	1	2	3	4	5	6	7	8						
Dreidel	●	●	●	●	●	●								
Dreidel Decoration		●	●	●	●	●								
Magen David			●	●	●	●								
Tessellations					●	●	●	●						
Wind Rose				●	●	●	●							

SOUTH AMERICA

GRADES

ACTIVITY/GAME	1	2	3	4	5	6	7	8
Quipu	●	●	●	●	●	●		
Inca Bird Design					●	●		
Inca Duck			●	●	●			
Colombian Star			●	●	●	●	●	●

MIDDLE AMERICA

GRADES

ACTIVITY/GAME	1	2	3	4	5	6	7	8
The Maya				●	●	●	●	●
Maya Numerical Symbols			●	●	●	●	●	●
Patolli				●	●			
Aztec Calender			●	●	●			
Aztec Numerical Symbols			●	●	●			
Toma Todo	●	●	●	●	●	●	●	●

NORTH AMERICA

GRADES

ACTIVITY/GAME	1	2	3	4	5	6	7	8
Rain Bird	●	●	●	●	●	●		
Zia Sun Symbol				●	●	●	●	
Pueblo Fish		●	●	●	●	●		
Paiute Walnut Shell Game		●	●	●	●	●		
Tipis			●	●	●			
Eye Dazzler				●	●	●		
Hogan	●	●	●	●	●			
Navajo Fry Bread				●	●	●		
Window Rock Design						●	●	●
Wasco Deer	●	●					●	●
Quilts						●	●	●
Hex Signs								

NORTH AMERICA—THE HOPI

GRADES

ACTIVITY/GAME	1	2	3	4	5	6	7	8
Hopi Rain Cloud			●	●	●			
Pottery Design					●	●	●	●
Hopi Bird			●	●	●			
Sky Window						●	●	●
Sun Kachina Mask						●	●	●
Additional Activity Pages						●	●	●
Hemis Kachina						●	●	●
Totolospi				●	●	●	●	●